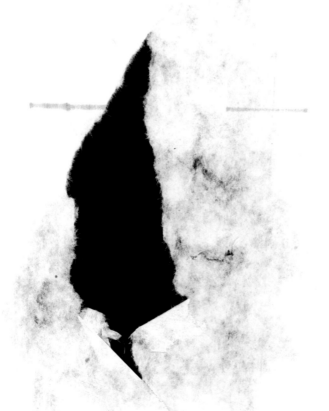

Rivers of
the Rockies

Rivers of
the Rockies

by Boyd Norton

RAND McNALLY & COMPANY

Chicago · New York · San Francisco

For Jean Anne and Scott,
with the fervent hope that there will
always be wild rivers
for them.

ACKNOWLEDGMENTS

Wish I could take exclusive credit for this book. Unfortunately I can't. Too many people contributed in too many ways to make this what it is. In particular, I want to thank:

Cort Conley and Jim Campbell of Wild Rivers Idaho, for their help in reviewing portions of the manuscript and aiding in my historical research.

M. Lizbeth Carr, for providing me with valuable information on John Wesley Powell.

Lynn Taylor of Denver Public Library and Terry Mangan of The State Historical Society of Colorado, for their aid in helping me find many of the historical photos.

Finally, I must thank my wife, Barbara, for reviewing the manuscript and setting me straight from time to time.

Book Design by MÀRIO PAGLIAI

Copyright © 1975 by Rand McNally & Company
All rights reserved
Library of Congress Catalog Card Number: 75-7798
Printed in the United States of America
by Rand McNally & Company

ISBN 0-528-81012-X

First printing, 1975

Overleaf: Pilot Peak and valley
of Clark's Fork of the
Yellowstone River, Wyoming

Contents

Fossil Mountain, western Tetons, Wyoming

Introduction

I MADE MY FIRST RIVER TRIP almost 20 years ago. (My God, has it been *that* long?) No big hell-roaring adventure of white water, it was on the placid Little Huron River in Michigan's Upper Peninsula. There were three of us in a canoe and with enough beer to float us out of there on the empties alone. Those two brief and quiet days were a welcome respite from classes in differential equations and advanced calculus and nuclear physics at Michigan Tech. There was nothing but us and the river and the canoe (and the beer, of course). We swam, talked, fell in the river, drank beer, hiked some of the forest trails, and not necessarily in that or any other logical order. In turn we got lost, rained on, bitten by mosquitoes, and sunburned. We cooked on an open campfire, solved the world's problems with our profound discussions, and listened from our sleeping bags that night for the wolves purported to be in the area. (We didn't hear them.) In short, it was a memorable trip. Now, 20 years and a few thousand river-miles later, I still get sweaty palms and a racing pulse each time I push off from shore and out into the current.

Rivers. Something about 'em. Almost like railroads in a way. Strange analogy? Not really. Rivers, like railroads, have been endowed through song and poetry with the mystique of adventure. Movin' on. Seeing what's around the far bend. Call it the Mark Twain syndrome if you will, but it's real. Even today, on rivers I've run many times before, I find myself getting restless sitting on the shore watching the swift water roll by.

But—there always has to be a "but" somewhere—too many of our rivers have lost their allure and beauty. I recall a few years ago an article recounting a journey on the Concord and Merrimack rivers, retracing Thoreau's famous journey. Only it was a horror show of stench and filth and garbage compared to Henry David's trip. And a friend of mine tells of new challenges in river running, like avoiding typhus if you get dunked in the Potomac. And then the oft told stories of the inflammable Cuyahoga or the polluted Hudson or the . . . Well, it's a long list, and unfortunately the stories are true. Too many of our rivers have become objects of commerce and contempt, dammed and damned. I've been afraid to go back and revisit the Little Huron River, afraid of what "they" have done to it—"they" being all those nameless, faceless entities responsible for changing and lousing things up.

And so I live here in the Rockies, the last stronghold. Stronghold for what? Well, rivers for one thing. True, some nasty things have happened to the rivers even here. By and large, though, they still run pure, free, and wild in the Rocky Mountain country. The way they ought to.

. . . .

Rivers are born of mountains. Big mountains spawn big rivers. Here in the Rockies one of the biggest rivers of the continent flows by my back door, though most people don't believe me when I tell them so.

Alongside my home, located at 8,500 feet near the town of Evergreen in the Front Range of Colorado, flows a stream. It is not large, perhaps a yard wide in most places. It begins in a little valley a mile or two up the road. As I look out my window the stream is dormant now, covered with two to three feet of dazzling snow and chilled to quiescence by 10-degree temperatures. In a few months it will come to life, carrying the melted snow from the high hills around us. And in the summer we will again sit in the long grass next to it, surrounded by wildflowers and shaded by aspen and willows, listening to it tumble by.

I like to tell our occasional guests that this stream (I've never found out if it has a name) is the source of the Mississippi River. That, of course, never fails to evoke some surprised and sometimes vehement denials, especially among those who were taught in grade school geography that the source of the Mississippi was in Minnesota. But I persist, pointing out that a few hundred yards down the road our stream joins a larger one, called Cub Creek. And, for those not familiar with our local geography, Cub Creek wends its way between these hills to join, about six miles farther on, Bear Creek (what else?). From Evergreen, Bear Creek begins a rather steep plunge down the eastern flank of the Front Range to the flat plains below, a total drop of nearly 3,000 feet. Near Denver, in what we mountain folk refer to as "them flatlands," Bear Creek joins the South Platte River. The South Platte merges with the North Platte, becoming simply the Platte, which in turn joins the Missouri, which empties into the Mississippi. The waters flowing by our feet here will eventually reach the Gulf of Mexico after a journey of more than 3,000 miles.

My argument is designed, of course, more to dramatize than to convince. The logic could easily apply to any one of a hundred streams nearby or any one of the thousands of streams emanating from the Rocky Mountains, each of them the source of the Colorado or the Missouri or the Columbia or the Rio Grande or the Mississippi, and so on. This dramatization does illustrate, however, the vast and subtle influence exerted by the streams and rivers of the Rockies. We tend to forget that the Rocky Mountains, though they lie more than two-thirds of the way across the continent from the Atlantic Ocean, affect both directly and indirectly the land features, cultural patterns, and climate of a large part of this country.

Rivers of the Rockies. Is there a typical one? Does my little stream, tributary of the South Platte, and indirectly of the Missouri and Mississippi, represent a typical situation? Yes and no. Trying to discover a typical river is like trying to find the average man. There may be some superficial similarities, but in the long run each is unique. For example, just over the hill from my home and about 30 miles away as the crow flies, the streams are all tributary to the Colorado River. In contrast to the South Platte-Missouri-Mississippi system, which meanders across the Great Plains country, the Colorado network has carved a vast system of canyons in the bare and colorful slickrock country of Utah and Arizona. In further contrast is the Snake, born of Yellowstone and the Tetons, flowing through the dark and mysterious lava country of Idaho. Or there's the Salmon in its wilderness vastness. The rivers of the Rockies, it seems, are typified by diversity. And in that diversity are found some of the most amazing and beautiful land features of the continent.

The Rockies are big mountains, both in vertical height and in geographic scope. Extending from northwest Canada nearly into Mexico, this cordillera is made up of a number of separate but ecologically connected mountain ranges. And though most of these ranges are separated by prairie or desert, there seems to be one connecting link, a common thread that ties them together: water. Being landbound, we lose the overall perspective and tend to forget the grand scale of it all, as demonstrated by the case of my backyard stream. But from an astronaut's prospect a hundred miles or so above it, this backbone of the continent seems flattened by distance and perspective. And an orbital view reveals the most distinctive feature of the earth's landforms: the work of water. Etched deeply into that wrinkled surface is a delicate lacework of streams and rivers, a vast web of water that drains and scours and shapes the land.

The origins of such intricate patterns are humble. Drops become trickles, trickles become rivulets, rivulets become streams, and streams become the rivers which carry these collected waters in a final, unified race to the sea. The storms that ultimately feed the rivers transport oceanic moisture back overland, across the Sierra Nevada, the Cascades, across the dry basins of Nevada and eastern Oregon and Washington, and sometimes up from the Gulf of Mexico. Then the storms drop their moisture on the high peaks of the Bitterroots and San Juans and Uncompahgres, the Sangre de Cristos, the Tetons, Bighorns, Lemhis, Wasatch, and a host of other ranges. Fed by glacier or melting snow or the raindrop dripping from a spruce needle, these waters fight their way back to the sea again to close the cycle.

The cycle began eons ago when the land was flat and featureless. There were no mountains. And there were no rivers. There was plenty of water, however, for a vast ocean once covered the Rocky Mountain region. Out of this ocean were precipitated the sediments and skeletons of marine life that later became sandstones and limestones, some of the building blocks for the mountains to come. In time the ocean receded as the continental land mass beneath uplifted slowly. For countless millennia the land was cloaked with lush vegetation, and dinosaurs roamed the forest and plain. Then, with subtle slowness, the sea invaded again. And again receded. All of this over a time span of hundreds of millions of years. But even before that last era of seas had ended, the restless earth had begun folding and faulting, and the once flat plains were buckled into hills, forerunners of the mountains to come. Beginning some 70 million years ago and ranging in distance from Alaska to Mexico, stresses built within the plastic mantle of the earth, and earthquakes rumbled across the land. The Rockies were not born as a single upthrust of mountains, however. Each of the ranges grew as a separate entity, inch by inch over millennia, and each over a vastly different time scale. Among the earliest ranges to rise above the land was the Front Range of Colorado, the very hills and peaks that surround my house. This uplift began about 70 million years ago. And among the youngest of the Rocky Mountains are the granite spires of the Teton Range, whose spectacular peaks began rising above the flat valley of Jackson Hole a mere 9 million years ago.

Where and when were the first rivers born? Incipient rivers were being formed before there were any significant hills or mountains. At any place where the land had a slope or gradient, the force of gravity drew the accumulated waters relentlessly back to the sea. At first these streams scoured the soft earth; then they began to work on the harder bedrock beneath. As higher hills and mountains took form and shape, the associated streams and rivers were altered accordingly. Steeper slopes meant swifter currents, which in turn meant faster cutting action. But there were more subtle effects as well.

Moisture-laden winds from the ocean blew inland across flat plains, and occasional storms fed the land with life-giving moisture. In time, however, the growing mountain ranges became barriers to the moving storms, impeding their flow. Forced upward by the high ranges, these storms were chilled to release even more of their moisture. More streams were formed, swift mountain streams that cut and carved the land. The soaring mountain ranges created the very forces that would tear them down, the streams becoming a built-in mechanism for self-destruction.

The glacial epoch was relatively recent in terms of the total geologic time span of the Rockies, but the ice ages had a profound effect, shaping land features and, in many cases, altering the rivers of the Rockies.

Beginning about 200,000 years ago, climatic changes brought about a time when the summer sun no longer melted off the accumulation of winter's snows. Winters grew longer and colder. Under such conditions the ever-increasing layers of snow were compacted under their own weight and transformed into ice. Great rivers of ice flowed downward from the lofty heights of the central Rockies and spilled out into the nearby plains. In the northern reaches of the Rocky Mountains, the land was covered with a great ice sheet that stretched across the continent. Only the tops of the higher mountains rose above this sea of ice, a frozen landscape similar to that of the Antarctic or Greenland today.

For whatever mysterious reasons, the cycle turned and a gradual warming took place.

Thunderheads over western Wyoming

Snake River, Swan Valley, **Idaho**

Slowly the continental ice sheet melted and receded, and the rivers of ice retreated into the mountain fastness, leaving the land free again of ice but greatly altered. Great peaks had been chipped and scoured, sharpened and faceted by the ice. Canyons were broadened, and in the valleys were deposited the debris of the receding glaciers—the scrapings of rock, ground and polished, from the mountain ramparts. In some places river courses had been changed. But perhaps more important, the massive accumulation of snow and ice, as it melted, fed new streams which cut new channels. With such a huge volume of runoff, one can well imagine that both the new streams and the older river courses carried immense amounts of water, altering the land still further. In time much of the land recovered, and grasslands and forests concealed the naked remains of glaciation. But it wasn't to last.

The second glacial epoch probably began about 80,000 years ago and lasted some 50,000 years. Once again the mountains gave birth to rivers of ice that crawled forth to rule the land. And in the north another ice sheet covered large areas of land, though not as great an expanse as the first. As we shall see, this second ice age played the key role in man's migration to North America. This epoch also passed and gave way to warmer times and new land features.

The third and most recent of the ice ages began only some 13,000 years ago and lasted about 5,000 years. Although it was the mildest of the glacial periods, it too had an effect on the land. Here and there along the chain of the Rockies today are found the remnants of those once mighty glaciers, ice fields now tucked away in the remote and shady recesses of the mountains as though waiting for the cycle to turn so that they may once again creep outward.

Rivers obviously are not static features. But we tend to think of river courses as being fixed, at least in terms of our own brief life spans. In the larger scope of geologic time, however, whole river systems have changed radically, some disappearing entirely.

Like people, rivers begin young and vigorous. Given time and the right conditions, they can carve great gorges like the Grand Canyon or Hells Canyon. But in time they also mature and wane in their sculpting chore, broadening the steep chasms they have created, as though dissatisfied with their initial work. And given even more time, these same rivers reach old age, broadening their valleys still more by meandering back and forth, becoming sleek and fat and placid, their hardest work done. Whole landscapes have been altered by this work of rivers; whole mountain ranges stripped and torn down and transported away grain by grain.

At this point in geologic time, all the various stages of river development are present in the Rocky Mountains: old age, maturity, and especially youth and vigor. The streams and rivers here have been shaped by this land and they, in turn, have shaped the land itself. The result is an amazing region, perhaps the most varied and beautiful on the continent. What other place can boast the four deepest river gorges in the world? Or a geography so varied as to contain rugged snowcapped mountain, cool alpine lake, searing desert, open prairie; vegetation ranging from saguaro and ocotillo to lodgepole and ponderosa to lush white pine and cedar? Or a climate that offers temperature extremes from 50 degrees below zero to 120 above? Where is the air cleaner, the land wilder, the water purer? And in how many other regions can you hear an elk bugle, catch a 30-pound salmon, float a raging river, fall off a 14,000-foot mountain, ski waist-deep powder snow, get bitten by a rattlesnake, see a cougar, smell the perfumed air of sage after a rain, bake in sun or freeze on a glacier —and sometimes all of these in a single day?

Well, not many. Thus what's contained here in word and picture is a small and random measure of the essence of all that. Hopefully. And if my feelings of pride, possessiveness, and concern for this country keep poking up through the text like pasqueflowers through a spring snow, then I hope you'll forgive me. Because I aim to keep it all just the way it is. If I can.

Headwaters

IT SEEMS THE ORIGIN OF MANY THINGS, this land above timberline. Headwaters? Yes, it is the gathering place of melting snows and trickles of water that coalesce from the mountaintops into the streams that become our rivers. But such things as soil also have their origins in this bare rock, bedrock land in the sky. For here the chippings and scourings, erosion and corrosion of water and ice imperceptibly reduce these massive ribs of planet earth into miniscule particles to be transported by rushing water to the lowlands and valleys, there to be deposited as the soft, life-giving mantle of soil. And so one might say that much of life itself has its origins here in this harsh land.

Life. The higher one goes in the Rocky Mountains, the more one realizes how thin a mantle harbors life on earth. I am lying on a spongy carpet of grass at an elevation of almost 12,000 feet in the Colorado Rockies, and at first glance life seems pretty sparse here. This morning I left car and highway behind me and began my hike in the coolness of a spruce forest where an incredible variety and diversity of life forms surrounded me. A deer bounded across the trail shortly after I started out. Flowers caught my eye next to the stream: Indian paintbrush, columbine, geranium. Even a rare calypso orchid in one place. Birds, chipmunks, insects. And a squirrel who objected loudly to my intrusion. From the microscopic to the macroscopic, there were seemingly an infinite number of living things around me, things that I not only could see, but hear and smell as well. Now, four hours later, it seems that I am approaching the upper limit for survival of living things.

My original goal for today had been to seek out the origins of the Colorado River high among the peaks of Rocky Mountain National Park. But right from the start I suspected that my journey would be a futile one. As the forest thinned and I began to emerge from this coniferous zone, passing the demarcation known as timberline, I found my original stream reduced to a mere trickle in places. Moreover, if I chose, I could have traced it back to a hundred little origins: a melting snowbank here, a pool of placid water there, or a filmy glaze of moisture on a smooth outcrop of granite somewhere else. To put a label on any of these, I discovered, and say that this or that place is the *beginning*, the very start of the Colorado River, was meaningless. Carrying it to a logical and absurd extreme, I might also search for the single molecule from which the great river springs and begins its long journey. And so, with my original goal reduced now to an exercise in semantics, I am engaged in the casual study of the alpine tundra landscape that stretches for miles around me.

Tundra. The very sound of it seems to ring with the booming of glaciers or the violence of mountain thunderstorms. And while it is true that tundra is treeless, it is far from devoid of life. As I lie here watching white clouds race across the blue-black sky overhead, 13

Big Southern Butte, Idaho

I discover an amazing number of miniscule flowers, all within reach of my outstretched arm.

One is tempted to describe life here above timberline as primitive, but to do so would be a gross error. In fact, quite the opposite is true. Most of the plants and animals of the alpine tundra have evolved rather sophisticated and complicated ways of coping with this harsh climate. Take, for example, the little alpine buttercups that grow in a cluster near where I am lying. A biologist friend once told me that these delicate yellow flowers are several years in the making. For the first few years of its life, the plant struggles to survive, drawing nourishment first from its microscopic seed pod and later from the thin and sparse soil. Its growth processes, though slowed by the long, cold winters, may begin with only the slightest warming, only a degree or two above freezing, thus taking full advantage of the all-too-brief warmth of summer. At a certain stage of its development, the tiny plant begins focusing its energies toward producing a small cluster of cells that represent the embryonic flower. This may begin one summer and continue the next, with a pause for dormancy in winter. Finally, after four or maybe even five summers in the making, the newly formed buds burst open to drink in that brief warm sunlight and to attract those insects that will pollinate the plant and continue the cycle. It seems an exercise in patience, if one may attribute such traits to a plant.

Another slow grower is the moss campion, a spongy green plant cluster with brilliant magenta flowers. Again according to my biologist friend, this clump of flowers nearby may have taken as long as 25 years to reach this stage of showy blooms. I feel the need to step a little more carefully in order to avoid crushing any of this delicate beauty that was so long in the making.

My own favorite is the sky pilot, also known as the alpine skunkflower, an ignoble name for such a pretty thing. Admittedly this flower is hardly what you would call fragrant, and the crushed leaves of the plant do have a decided skunklike odor. But I remember one time sitting on a bare granite ledge high in the Tetons, waiting out a storm on a climbing trip. I was cold, wet, miserable, and, frankly, scared as hell as the storm raged around me and my companions. While moving into a more comfortable position, I felt something brush my hand and discovered the most delicate little blue flower growing in what could barely be called soil; more like pulverized granite trapped between two big blocks of rock. The flower waved violently in the ferocious wind that assaulted us, and pellets of sleet were bombarding it unmercifully. But it hung on, for it had survived a long time in this world of snow, sky, and granite. And it would probably do so for a long time to come. The storm passed on as all storms must, and we moved on to finish the climb. But I've remembered ever since how cheered I was by that little bit of beauty hanging on in what has to be the virtual upper limit of survival.

Right now I am forced to move on by another of the tundra's life forms, one which I have a bit of difficulty appreciating. The mosquitoes here seem to be carnivorous, and one spends a lot of time swatting and shooing. They thrive on most insect repellents. But just as well, I think; I have been lolling in the warm sun here for too long. I must continue my Quixotic quest for the source of the Colorado River. Perhaps up there, higher on the ridge, I may find that mysterious wellspring.

Flowers are everywhere as I walk across the soft and spongy carpet of grasses and sedges that make up the tundra. Alpine forget-me-nots, more moss campion, saxifrage, more buttercups, rockjasmine, and a dwarf version of something that looks like the fleabane that grows near my home in Evergreen. There are a great many more that I can't identify. My view is superb as I hike along. The peaks of the Never Summer Range glisten in morning sunlight. Stretching around me is a great panorama of mountains, all of which are devoid of any obvious vegetation and capped with bare rock and snow. Timberline, that tenuous demarcation between forest and tundra, wavers across the flanks of the mountains. Far from being constant, timberline varies a great deal from place to place. In this part of the Rockies, it averages around 11,000 feet; but farther north, in the Tetons or the Sawtooth Range, for example, it may be as low as 9,000 feet. And south of here it may be 12,000 feet or higher.

Farther up the ridge I find some patches of fresh snow in the shadows of boulders, the residue, apparently, of a recent storm. Snow? In late July? Summer as we normally think of it is really a theoretical abstraction here on the alpine tundra, for summer comes late to this country. If at all. Reminds me of the joke about the weather around my home, a mere 8,000 feet above sea level. "Only two seasons in Evergreen," the saying goes. "Winter and fourth of July." And, if we're lucky, it doesn't snow on the fourth of July.

Up near the ridge crest, the tundra is no longer a smooth carpet of plants. The land is broken here, and piles of granite slabs and boulders lie about. But life hangs on even here, for nearly all the exposed granite is coated with patches of lichens. This symbiotic combination of alga and fungus comes in an amazing variety of psychedelic reds, greens, and yellows. More common is a gray-green color. And there are blacks and browns as well. There is hardly a boulder or exposed piece of bedrock that isn't covered with the crusty and colorful patches.

The lowly lichens play a rather sophisticated role in alpine ecology. They are the soil producers. At least, they contribute a great deal to the production of soil. Over an incredible span of time, the rocks to which they are affixed are broken down by tiny secretions of acid from the lichens. The work of the acid is aided by frost and water, and the rocks gradually weaken and disintegrate grain by grain, forming the coarse soil of the tundra, part of which is slowly washed down in the streams to the lowlands below. The primitive lichens, then, become one more link in the life chain of the alpine tundra that contributes to the life of the temperate and moderate climes below.

Reaching the top of the ridge I discover, and not surprisingly, that there is no Great Source, no hidden lake of the clouds that gives birth to the great river. Instead there are more patches of snow, broken, lichen-covered rocks, and many more flowers that bloom in the interstices of the rock-strewn slope. And another of the tundra's life forms, the pika, squeals at me from behind a boulder. A warning? A complaint? Hard to say. On occasion when I've been hidden from their sight and posed no real or imagined threat, I've heard them squeaking merrily away, seemingly just for the sheer joy of it. The pika, or alpine haymaker, could easily be the model for some of those soft and cuddly children's toys. It is a gray-colored ball of fur with short ears and legs and essentially no tail, a sort of compact model of the rabbit. In fact, it is a lagomorph, that order that includes rabbits. The pika is well adapted to this bitter cold country, for its short ears and legs reduce heat losses. Its home is the boulder fields of the tundra, and it makes a living by gathering plants and grasses to store for the winter since it does not hibernate. But before storing those plants it spreads them out to dry on a sun-warmed rock, much as a rancher cures his hay before baling and storing to prevent its spoilage by mold. I can well imagine that the pika is quite comfortable in its winter den, munching on its stored food while a perpetual blizzard rages outside. But again I'm puzzled by those strange quirks of nature: Why doesn't the pika hibernate? Being snowbound for eight, maybe even nine months of the year must bring on a hell of a case of cabin fever. And the brief weeks of relatively warm weather must be spent gathering grasses to tide it over until next year. Hardly seems fair.

• • • •

As a mountain freak I've spent a lot of time above timberline in this alpine country. It's inevitable, I guess, that if you hang around mountains long enough, sooner or later you'll be dumb enough to climb one of them. Not all of them are the ropes-and-pitons kind of climbing, of course. There are many places in the Rockies where peaks—some of them quite high and spectacular—can be strolled up. What my hard-core mountain climbing friends refer to as a "scramble." But even these easy hikes are not to be taken lightly, for while many plants and animals have adapted to this frigid land, man is much more vulnerable when the weather turns bad. Many people have died in the alpine country because they were ill-prepared for the vagaries of weather. As I hike along the ridge, I keep a sharp eye out toward the west, watching for signs of those harmless cumulus clouds transforming into the forbidding and dangerous cumulonimbus that mean storm.

Snake River in Jackson Hole,
Grand Teton National Park,
Wyoming

Aspen leaves

My introduction to Rocky Mountain storms occurred many years ago on a climb of Middle Teton in Grand Teton National Park. In fact, it was my very first climb in the Teton Range. And damned near my last. We had started before dawn from base camp, climbing steadily up the ever-steepening snowfields. On reaching the saddle between the Middle and South Tetons, we were able to view the western skies for the first time that morning, and what we saw was frightening. Moving toward us from the flat Idaho plains was a series of dark, almost black thunderclouds from which jagged fingers of lightning reached out to touch the ground with a delicate grace. We mistakenly assumed that we could outrun the storm—it seemed miles away—and reach the summit and retreat before it hit. But we found ourselves some 500 feet below the peak when the first dark clouds enveloped us. For some unknown reason we continued on rather than retreating and for what seemed like hours sat

Hidden Cascade
on Rock Creek,
Elk Mountains, Colorado.
Photograph by
William Henry Jackson,
1874.

huddled among the gray blocks of granite on the summit while fierce winds hurled sleet into our faces like buckshot. Several times the blocks we sat on hummed and crackled with electricity. Finally coming to our senses, we beat a hasty retreat back down the steep—and by now icy—ledges, down the slick snowfields, rappeling in places where hand- and footholds proved to be too treacherous. We made it back to base camp safely, mindful of the fact that we had been extremely lucky. Such storms have claimed a lot of victims.

There is an awesome power to Rocky Mountain storms. Since my first encounter in the Tetons, I've spent many an anxious hour huddled somewhere near or above timberline, counting off that interval between the flash of lightning and the crash of thunder as a storm passes through. And on a few occasions there was no separation in time between the blinding arc and the ground-shaking explosion. I have an overpowering desire to be elsewhere at times like that, and there are vows never to return ever again to this damnable country. But then, we mountain folk are all a little crazy, I guess. We keep coming back.

•   •   •   •

Now I must give up my sojourn, my search for the elusive river source. Clouds to the west are beginning to mass, gathering themselves in great convoluted forms that warn of violence. I must begin my descent.

It's with reluctance that I start back down, hopping over the boulder field, stepping across the patches of snow still glaring white in the sunlight, down across the spongy turf of tundra grasses and wildflowers, following those cold clear trickles of water as they race downhill toward Utah and Arizona and the Gulf of California. The beauty of the tundra is much like that of the desert. There's an openness and feeling of freedom about it. And a simplicity in the life and landforms. It is the atmosphere of purity that appeals to me most, however, for it is a land still largely unaffected by man. Like the desert it has an intensity of climate, a harshness that places severe constraints on man's activities. You can't farm it, graze it, or log it. Fortunately. Living here becomes a struggle for survival. But armed with technology we are making inroads into this headwaters country: roads and highways, large mining operations, growing numbers of recreational toys, and even weather modification schemes. In his novel, *Wellspring*, Edward Hawkins points out with shocking clarity the vulnerability of our society to tamperings with our life-giving waters. In that story, at a single spot in the Colorado Rockies a deadly biological agent is released simultaneously into the headwaters of the Platte, the Arkansas, and the Colorado rivers, subsequently affecting millions of people. While the story is fictional, the chilling overtones cannot be ignored.

There is a paradox about the tundra, for though the life forms are highly adapted to fierce and hostile conditions, they are also very fragile, very sensitive to changing conditions. This land heals with infinite slowness. A road scar may take centuries to recover, while in milder and moister climes roads can be overgrown in a tangle of vegetation in a mere decade or two.

As I wonder about our ecological meddlings I once more pass timberline, plunging downhill past some gnarled and windsculpted spruce and on into the thickening forest. Heading for the faint trail, I parallel the stream I followed earlier, on my ascent. The sky is dark now, and the wind is chilled. In a few more miles light rain is falling, and I stop to pull a rain poncho out of my rucksack. As I continue down the steep trail, the sound of my clumping boots is interrupted every few moments by the rumble of thunder up above. I watch the rain coalesce into glistening drops on the needles of Englemann and blue spruce as I head back to my car and civilization, thankful that I didn't tarry too long up there on the tundra where it is probably snowing again. The rain has released the deep, earthy smells of the forest, the resinous odor of evergreens, the musky smell of damp earth and decaying humus, the rich and sensuous odor of life. It's a welcome feeling to be back in more familiar and friendly surroundings. Perhaps we need to be exposed to some austerity and harshness from time to time. Makes us appreciate the amenities all that much more, I think to myself, headed back down with thoughts of a blazing campfire, a big meal, and a warm sleeping bag.

Cascade Canyon,
Grand Teton
National Park,
Wyoming

Teton Glacier,
Grand Teton
National Park,
Wyoming

Western Tetons, Wyoming

Early Man in
the Rockies

IN MY TRAVELS ALONG the river byways of the Rocky Mountains, I've made some interesting and sometimes surprising discoveries about the relationship of man and rivers. Most readily apparent today is the effect of man upon the rivers—his irrigation channels, his huge concrete dams. What is not so apparent are the ways in which rivers have influenced men, the important role of rivers in shaping, directly and indirectly, the various cultures of man that emerged on the North American continent centuries before Europeans rediscovered this land.

The history of man in the Rockies is inextricably tied to the history of rivers in this region. The story is not easy to tell, however, for it is like a vast jigsaw puzzle in which many of the important pieces are missing. Direct evidence of early man is not readily seen today, much of it having been masked by the work of time and, in some cases, by the works of our own culture. Most often that evidence has been fragmentary and well hidden anyway, tucked away in the wild recesses of the land. I remember a trip into the wilderness of Escalante Canyon a few years ago in which I took off to explore a side passage, following its twisting course deeper and deeper into unknown slickrock country. There was no evidence of anyone ever having been there before, and I rather smugly assumed that I was the first human to set foot in that part of the desert wilderness. Then I rounded a bend and discovered the strange petroglyph of a snake scratched into the smooth, red rock of an alcove. Someone had passed that way before, a couple of centuries or perhaps a couple of millennia ago, and left—what? A warning? A welcome? Then there was that rounded boulder with its cryptic figures and signs which I found sitting in the tall grass of a meadow several hundred yards from the Snake River in Hells Canyon. And the figures in a cave next to the Middle Fork of the Salmon. Or that incredible panel of petroglyphs in Grand Gulch near the San Juan River.

But these encounters with prehistoric man have all been relatively brief, pauses here and there to gape in awe or take a picture, then move on. I've had a compulsion for years to spend some time in one of those dwelling spots of ancient man, watch the sun rise and set, feel the changing moods of the day, and try to capture, in some small and personal way, a glimpse of the lives of these early people. Finally, on a recent trip to Mesa Verde National Park in Colorado, the opportunity presented itself. Hiking a nature trail out to a canyon overlook, I crossed a small gulley, a dry sandstone wash that looked like it might be the beginning of a steep canyon. I left the trail and followed the wash, much against National Park Service rules (unlike other national parks, Mesa Verde restricts back country hiking in order to preserve its ancient ruins). I rationalized my illegal venture on the ground that I would take special pains to see that I disturbed nothing should I find any artifacts or ruins. Sure enough, that wash soon became a steep gulley, then a small canyon carved into solid rock. As I clambered over a series of drop-offs, heading down into the deeper recesses of my little canyon, I spied the walled ruins of an ancient structure under the eaves of a cave on the east side of the gorge. Since it was considerably off the beaten path of the regular tours at Mesa Verde, it occurred to me that there, at last, was my own private little ruin where I could sit and contemplate the ways of a thousand-year-old culture. Getting to that cave proved to be a bit more difficult than I had anticipated, however, involving some tricky rock climbing. I was tempted to head back to the car, stuff my sleeping bag and some cheese and raisins and nuts into my rucksack, and come back to spend the night in the cave. I decided against it, mainly because I was too lazy and the hour was late.

So I did the next best thing. After watching the sun set beyond the far rim of the canyon and listening to the yapping of some coyotes somewhere, I headed back to the car in deepening twilight with the vow of returning next morning. After a windy night in the Mesa Verde campground, I set out before dawn to keep my appointment with long-departed canyon dwellers.

. . . .

The sunlight plays strange tricks on a crisp autumn morning in these canyons. As the first rays of light slice across the mesa top and gild the upper edges of the cliffs, I get a feeling that something is moving nearby. Perhaps it is nothing more than a mule deer slipping quietly along the fringe of piñon pine and scrub oak, coming out of the stunted and

Canyon De Chelly, Arizona. Photograph by Edward Curtis, ca. 1885.

Water ouzel

Middle Teton, Grand Teton National Park, Wyoming

shadowy forest to bask in the warming rays of light. Or it may be the movement of the forest itself in the barely perceptible wind. The wind. It too plays tricks on the senses as it whistles along canyon walls, moans through these crumbling ruins long abandoned. There is —I almost hate to use the term, but I must—there is a haunting quality to the sound of the wind here, aided by the echoes and acoustics of soaring cliffs, dead-end canyons, curving alcoves. There is an eeriness, even now in broad daylight, that brings reminiscences of childhood fears of haunted houses and the unknown darkness. Sitting here amid ruins of an ancient structure built more than a thousand years ago, I sense ghosts all about me. With the first light of dawn, I begin hearing the distant murmur of voices from a village across the canyon: men calling to one another as they prepare to climb the steps carved into the cliff to tend the mesa-top fields of crops; children laughing and playing in the morning sunlight on the flat floor of their arching cave; women chatting as they busily grind corn or mend baskets or make their traditional clay pots. The sounds are there; the people are not.

Ruins at Canyon
De Chelly, Arizona.
Photograph by
Timothy H. O'Sullivan,
ca. 1873.

It was the Spanish explorers who gave this place its name—*Mesa Verde,* Green Tableland—when they passed near here in the late 1700s. It is a strange, truncated, mountainous mass lying at the interface of the Rocky Mountains and the great Southwest Desert, creased with numerous canyons like the one in which I now sit. Carved by many little intermittent streams that empty into the Mancos River, these canyons are steep and rugged. Over the ages this sandstone has weathered into a lovely, feminine softness. Edges have been smoothed by time, wind, and water. Vertical cliffs have eroded into gentle alcoves and caves, giving the appearance of having been sculpted by hand into free-form shapes. Despite these softening features, however, there is still a harshness to it all. It is an arid place. The ground bristles with yucca and cactus, juniper and scrub oak and piñon pine. The breeze of October has a sting that portends a bitter cold in winter. It seems an unlikely place for people to have lived, but starting nearly 2,000 years ago a great many did. Mesa Verde is one of the more fascinating stories of early man on our continent.

The sun has another two to three hours' worth of travel before reaching the outermost lip of my niche in this east wall of the canyon. Even with a down parka, there is a bone-chilling cold to these shadowy recesses. The rock itself is numbing to the touch. Perhaps it was around warming fires on mornings such as this that the people recounted in chant and song the legends of their own origins. Is it possible that some thread of the factual story lasted through the 40 millennia that had passed since the arrival of man in North America? Or did the struggle of 40,000 years of migration and survival erase all memories of man's ties to the Asian continent?

The evidence is now strong that nomadic hunting tribes crossed the Bering Land Bridge sometime during the second of the three great ice ages of North America. It would seem an unlikely time for Arctic travel, but when vast amounts of oceanic moisture became locked in the ice caps on the northern edges of the continent, the water level dropped sufficiently to expose a land connection from Asia to Alaska. It was still a frigid land, similar to the frozen reaches of tundra that still exist in northern Canada and Alaska.

It would be wrong to assume that the first tribes to cross the Bering Land Bridge were bold adventurers, spurred on by visions of discovering new and glorious lands. In reality they were probably unwitting, and perhaps even reluctant, pioneers who happened to follow a new caribou migration path that took them far to the east of their traditional hunting grounds. Furthermore, when one lives on the very edge of survival in harsh Arctic lands, there is little time—or inclination—to indulge in such frivolous activities as exploration. No, it appears that it was survival, pure and simple, which motivated these people to follow the game herds across that frozen waste to inland Alaska and eventually, in scores of generations, to the interior of the continent. But from where I sit amid these Mesa Verde ruins, it is a long way in time and space from the crossing of the Bering Bridge. How did these people arrive here to build their fortresslike structures? And when? And why?

Well, explain my geologist friends, it seems that an interglacial corridor down the Mackenzie River Valley existed around the same time as the Bering Land Bridge. Thus, an easy crossing and an easy access to the interior of the continent existed together, give or take a millennium or two (geologists are rather extravagant with millennia; for my own part, I tend to be miserly with mere hours, days, and weeks, which is probably why I would not have made a good geologist). And so, over many generations, it was easy for the various tribes of this hunting culture to move southward along the eastern edge of the glacier-gripped Rockies until they reached the Great Plains. Here they found a warm, rich, and bountiful land, a place where the very earth shook with thundering herds of giant bison and where roamed camels and horses and giant mastodons, providing abundant food for them. Climatologists tell us that the polar ice caps had a great influence on the weather here and it was a far moister climate than exists today. The great herds thrived in the lush, expansive grasslands, and the people came to thrive as well.

But why move? If the land was so fertile, why would these people press on in search of new lands? Population pressures? Rivalry? Or, with abundant food, did man have the time to exercise that innate curiosity to see what lies beyond the next range of hills? Whatever the combination of forces, some did move on and ultimately roamed and settled not

Redfish Lake,
Sawtooth Mountains,
Idaho

only North America, but the whole of Central and South America as well. Of course, I make it sound as though it all happened in a lifetime or two. In reality, it was probably thousands of years before man had wandered from the Bering Straits to Tierra del Fuego.

But again the nagging question: Since man remained a relatively minor species on these continents, and therefore in no danger of depleting his food resources, what forces caused some to leave the bountiful plains and seek shelter amid the caves of this sparse and inhospitable land?

Tough question. As I look about me, the ruins offer strong evidence that these people feared someone, or some thing. This enclosure in which I am seated apparently offered defense against hostile intruders who might come up the steep slopes of the canyon below, and the overhanging lip of the cave prevented surprise attack from the rear. Whatever their fears, real or imagined, the people must have felt secure here. But, far removed from those bountiful plains, they had to make some profound cultural changes in order to survive.

While I have been pondering these questions, the sun has arced across the sky far enough to peer under the overhang of my cave. (Interesting observation: *my* cave. Having spent a mere few hours here, I'm already possessive about this place.) I saunter over to the lip of a low wall and seat myself in the warming rays of light. Suddenly the dank chill is chased away, and the day takes on a new aspect. The magic of solar warmth. Small wonder that many ancient peoples worshipped the sun.

The sun. As I bask in its warmth, it occurs to me that the sun may hold the key to some of those unanswered questions. Changes in climate could account for migration and cultural changes. We know that some 13,000 years ago North America was plunged into the grip of its third ice age. It lasted until about 9,000 years ago, and during the period when the frozen reaches of the Rockies and northern Canada were thawing out, the grasslands remained lush and fertile, the game herds plentiful. But when the swollen rivers had carried away the last of the glacial melt and prevailing winds were no longer saturated with moisture, many areas of the continent became drier. Much drier. Vegetation changed. In some places grasslands were replaced by desert. Game herds dwindled and tended to congregate around reliable water sources, becoming vulnerable to huge kill-offs practiced by the hunting cultures. In time, through both man and climate, certain species became extinct: the

Buffalo at water.
Photograph by
William Henry Jackson,
ca. 1868.

giant bison, the mastodon and mammoth, the sloth and camel and horse. (Horses would be reintroduced by the Spaniards in the 1500s.) With all of this in mind, it becomes a bit easier, then, to trace the transition in cultures. When game became scarce, man was forced to become a forager, supplementing his meager game supply with native berries and roots and nuts. Given enough time, such changes could logically lead to the development of agriculture.

Lounging in the warm sun here, I realize that I am guilty of that which I have accused my geologist friends of: extravagant wastefulness of millennia. Agriculture was not developed overnight. In fact, the earliest traceable origins of agriculture on this continent emanate from the sophisticated cultures of Central America, starting about 6,000 years ago. But it took nearly 2,000 years for this art to reach the Rocky Mountain region. Even then, it remained confined largely to the arid lands of the Southwest.

At first it seems a paradox that agriculture in North America would emerge first in the region least likely to support the growing of crops: the desert. But on closer examination, perhaps it is not so strange. This dry desert country of the Southwest was (and is) incapable of supporting herds of grazing animals, and thus the inhabitants of this region turned to other means of sustaining themselves: hunting small game and foraging certain native grains, fruits, and nuts. When the art of planting crops came to these people from Mexico, agriculture became a way of overcoming the sparseness of game in the Southwest. The planting of crops required water, however, and rather ingenious systems were developed to provide moisture and assure a reasonably successful harvest. As early as 2000 B.C. tribes in northern New Mexico grew a primitive type of corn not far removed from the diminutive (about the size of a man's thumb) wild maize, and it appears that they took advantage of natural seeps and springs to water the small plots they cultivated. In other areas "pot irrigation" was practiced, a scheme whereby people carried water in pots from a nearby source to the fields. Much later, around 100 B.C., the Hohokam people developed irrigation systems near the Salt and Gila rivers in Arizona. Using large ditches, and in some cases sizable earth dams to divert the rivers, the Hohokams transported water as far as 30 miles in a lacework of canals to nurture their crops. Once water was applied to the land, they found it amazingly fertile. Soon other people in the Southwest practiced such agriculture along river courses.

Wilting now under the hot rays of the early afternoon sun, I peel off my parka, stuff it in my rucksack, and decide to have a look around. At the northern edge of my cave I scramble across a slanting slope of rocks, hoping to find an easy route to the plateau above. A series of brushy ledges and a small gulley look promising, so I angle upward toward them. If this was the path used by the people of my cave, there's no indication of it. But then, it has been nearly 700 years since they departed, and such ephemeral evidence as trails would have been erased by wind, weather, and vegetation. After some bushwhacking, I find myself on the mesa top and I walk back along the rim to the approximate locale of my cave. Shuffling cautiously up to the curved cliff edge and leaning out as far as I dare, I'm able to see the rocky slope and the canyon bottom below, but there's no hint of the cave or the dwellings within. After a few moments I turn and walk way from the canyon and into the stunted forest of piñon pine and juniper. There's a resinous fragrance here, a mixture of sweetness and pungence. Picking up a piñon cone, I hope to find a few of the sweet and delicious nuts, but birds and squirrels have beaten me to them. Undoubtedly the cave dwellers made use of this plentiful food to vary their diet. Perhaps it was the chore of the children to gather piñon nuts while the men worked the fields and the women tended the cave dwellings and prepared the meals. There's no sign of any cultivated fields in this area, but again, 700 years could have erased many of the traces.

It was around the time of the birth of Christ that the Anasazi people moved to the Mesa Verde area. At first they lived on the mesa top, building shelters of wood and brush over shallow pits they had excavated. In winter they moved into the caves for relief from wind and snow, but they built no permanent shelters at that time. In moving here, these people brought with them their newly acquired art of planting crops and they found the soil rich and fertile. Irrigation as practiced by other people far to the southwest was out of the question here at Mesa Verde because there were few permanent streams to divert or dam.

Teton Creek,
headwaters of
Teton River,
Wyoming

Unnamed tributary,
Lochsa River,
Idaho

Upper Flathead River Valley, Mission Mountains, Montana

As it turned out, however, irrigation was unnecessary, for the nearly flat top of Mesa Verde averages some 8,000 feet above sea level, making it high enough to intercept moist air masses and cold enough to chill them and release precipitation. Snow is abundant in winter, and thunderstorms in July and August yield additional moisture. Summers are both hot enough and long enough to provide a good growing season.

Century after century the Mesa Verde culture flourished on this mesa top. In time, the people built shelters of rock faced with adobe or native clay, and these buildings were joined together in typical pueblo fashion. The added protection of stone and masonry against the weather allowed them to live year round on this plateau instead of seeking shelter in the canyon caves during winter. And the joining together of the domiciles gave added protection against outside raiders. Apparently the people were not secure enough, however, for around A.D. 1200 hostile nomadic tribes moved into the region and made frequent raids on these peaceful farmers. Under these pressures, the people sought the protection of the caves that had once given them merely seasonal shelter and they adapted their architectural techniques to these canyon alcoves. Although it moved them farther from their fields, they were much better protected against surprise attacks.

In fall this mesa top has lost much of its color of life. Even the perpetual verdancy of the evergreens is dulled. As I wander through the piñon and juniper, everything is dry and brittle. Grass stalks and seed pods of flowers snap and crunch underfoot. It is the time of dormancy, and everything is preparing for the siege of winter. I kick over a clump of grass, reach under it for a handful of soil. It is a dark, almost reddish, brown in color, coarse and granular and, because it has been such a long time since the last rainstorm of summer, it is also very dry. But come spring, this soil, supersaturated with melted snow, will nurture all the green plants for which this mesa was named.

The planting of crops by the Mesa Verde people was not undertaken in any haphazard sort of way. For generations the techniques and rituals had been honed to a fine-edged art. Though their methods may appear crude and their reasoning mystical, one must respect the farming skills of these people. After all, how many other civilizations can boast of nearly 13 centuries of longevity?

When the spring sun had melted the accumulation of snow off the plateau, the people began preparations for their annual crop. With the soil still moist from snowmelt, the land was cleared of weeds and the dirt loosened by sharp digging sticks. The family plots were not large, perhaps an acre or two in size, and they were scattered all across the plateau, interspersed with the piñon and juniper forest. In some cases the relatively flat stream courses at the heads of certain canyons were utilized for planting because the rains flowed along these channels and provided more moisture for crops. Occasionally rocks were piled to act as check dams on these stream courses, prolonging and regulating the water flow.

By May things were ready, but generations of experience had taught the people to heed certain signs, religious and physical, before placing a single seed in the ground. Priests were called upon to read and interpret those signs: the return of certain birds, the mark of optimum azimuth for the sun, the "feel" of daytime and nighttime temperature extremes. To plant too early was to risk killing frosts; too late, and the nourishing spring rains were missed. When finally the signs were right, the corn and squash and beans were planted in the rich, red soil.

In good years there were warm, gentle rains in May, followed by cloudless skies and a blazing sun in June. The sprouting plants grew vigorously, but by July they had sucked up all the stored moisture from the soil and were in need of water. In the normal pattern, July and August were the months of thunderstorms, and nearly every afternoon great billowing thunderclouds would form. These storms were brief, sometimes violent, but the moisture was soaked up by the growing crops. As summer gave way to early autumn, the afternoon storms diminished in number, and the crops matured under another cloudless blue canopy and a hot sun. Finally the time was at hand to harvest the bounty of earth, sky, and water.

Harvest was a time of rejoicing and celebration. Ears of corn and strips of squash were dried in the warm September sunlight, then carefully stored along with the beans and certain roots and nuts that had also been collected. The dry Mesa Verde climate allowed such easy storage, and there was little danger of spoilage. The best seeds of each of the crops were carefully set aside, perhaps with some religious ceremony, and these would serve for the next spring planting. When the deer and elk meat from hunting expeditions had been dried and stored, the people were prepared for winter, secure with an ample food supply and good-spirited with the knowledge that the gods had provided well for them.

There were, of course, bad years. When you've lived in the Rockies long enough you learn that in one out of every four or five years the winter is mild, snowfall light, and summers hot and dry. Just as frequently there is a long and bitter winter, with heavy snows beginning in early September and lasting until May. Here at Mesa Verde there might be a freakish storm where snow fell after the crops had been planted. Or those summer thunderstorms might smash the crops with hail instead of rain. Such minor vagaries of weather and climate the people could survive, for in good years it was frequently possible to store enough food to tide them over the bad. Through the more than 1,200 years that the people lived here, there were probably times of three, four, or maybe even five bad years in a row, and under those conditions life was undoubtedly grim. But always the cycle turned, the gods were appeased, and the bountiful years returned.

The Mesa Verde civilization lasted until nearly A.D. 1300, and its end was brought about by the very disaster that these people had lived in fear of for centuries: prolonged drought. Beginning in A.D. 1276 (a year pretty well determined by tree ring dating) there began a series of dry years that drastically reduced the crop yields. Snowfall was light in winter, and by the time planting took place the soil was already dry. Even worse, the spring and summer rains didn't come. This calamity of climate continued not for just one or two or three years. It lasted for more than two decades! It was a disaster of proportions that the people had never known before and it decimated their culture. As year after bad year went by, there was increased sickness and death as the people weakened. Toward the end there may have been a few cases of outright starvation. Certainly the very young and the very old suffered terribly. Undoubtedly they turned more to the hunting ways of their ancestors, but the plateau of Mesa Verde has never been rich in game, and that resource could have been drastically reduced, if not totally exhausted, in relatively few years. Ranging farther to the north, to the La Plata Mountains and the Mancos Valley, would perhaps yield more game, but it also exposed the people to the hostile nomads; and it was increasingly difficult to transport such food the long distances back to their home on the mesa top. In time there was only one solution to their desperate situation: They must move, leave their beloved home of countless generations, and find new lands.

It seems unlikely that it was a great mass exodus. Instead, small groups, perhaps a few families at a time, began to leave, probably starting after five or ten years of the prolonged drought. As more left, it became somewhat easier for those remaining to sustain themselves, with fewer mouths to feed. But by the tenth or fifteenth year of the drought, even the smaller numbers left behind found it difficult to subsist. People continued to leave. Finally, by the year 1300, Mesa Verde was totally abandoned, and the dry winds whistled through canyons and caves where once voices and laughter had echoed.

The drought affected not only the Mesa Verde people, but all the tribes of the Southwest in general. Inhabitants of Canyon De Chelly, Keet Seel, Betatakin, Hovenweep, and other desert areas were forced to leave their lands. Since drought was the cause of this migration, these people naturally sought lands where water was in greater abundance. They moved southward generally, settling along some of the more reliable river courses, such as the lower Colorado and the Gila. The Mesa Verde people apparently came to settle along the Rio Grande Valley in northern New Mexico and undoubtedly influenced the modern-day Zuñi and Hopi pueblo cultures.

Though the drought eventually abated in the early years of the 1300s, the people never returned to Mesa Verde. Their dwellings possibly provided nomadic tribes with temporary shelter from time to time, but they lay largely abandoned and undiscovered until the late 1800s, when Euro-Americans happened on them.

· · · ·

Castle Peak
and Frog Lake,
White Cloud Mountains,
Idaho

Beaver pond

The sun has been dropping in the western sky, and the air is becoming chilled again. I make my way back to the canyon lip and, after some poking around, I manage to find the obscure path back to my cave. As I arrive there, the cold shadow has crept up from the canyon bottom and is just entering the alcove. Once again encased in my parka, I sit and lean back against the wall of the ruin to watch the sun slip below the far canyon wall. A large bird circles in the air over the distant mesa top, but I can't make out from here whether it is a hawk or raven. Somnolence and silence creep in with the evening's twilight. Soon I must start back if I'm not to be caught by darkness, back to my car, the highway, towns, lights, other people, and all the trappings of my own culture. I wonder to myself what the Mesa Verde people would think of my society with all of its extravagant luxuries. I wonder also how well modern man might adapt to a more primitive life should some calamity befall our technological civilization. Perhaps not too well. Being a product of my own culture, I have no desire to give up the warmth and comfort of my home, the almost infinite mobility of modern transportation, highly efficient communication, antibiotics, abundant food, and a host of other things to which I have become accustomed. On the other hand, it would be rather smug of us to assume that happiness and a sense of well-being are states of mind unique to modern man. One tends to think of the lives of such people as the Mesa Verde inhabitants as being one terrible, joyless struggle for survival. Yet, with the exception of those last 25 years, we know that they lived well, were comfortable, celebrated the good times, mourned the bad, and undoubtedly took great pride in their accomplishments. Who can say whether they were any less happy than we? Or, to paraphrase Joseph Wood Krutch, Is happiness measured solely by cars or kilowatts?

As I clump over the trail back to my car in near darkness, some words of H. G. Wells come to mind:

> But in these plethoric times when there is too much coarse stuff for everybody and the struggle for life takes the form of competitive advertisement and the effort to fill your neighbor's eye, there is no urgent demand either for personal courage, sound nerves or stark beauty, we find ourselves by accident. Always before these times the bulk of the people did not overeat themselves, because they couldn't, whether they wanted to or not, and all but a very few were kept "fit" by unavoidable exercise and personal danger. Now, if only he pitch his standard low enough and keep free of pride, almost anyone can achieve a sort of excess. You can go through contemporary life fudging and evading, indulging and slacking, never really hungry nor frightened nor passionately stirred, your highest moment a mere sentimental orgasm, and your first real contact with primary and elemental necessities the sweat of your deathbed.

The Mountain Man

It is the time of fulfillment, the fullness of time, the moment lived for itself alone. The mountain men were a tough race, as many selective breeds of Americans have had to be; their courage, skill, and mastery of the conditions of their chosen life were absolute or they would not have been here. Nor would they have been here if they had not responded to the loveliness of the country and found . . . something precious beyond safety, gain, comfort, and family life.

Bernard DeVoto,
Across the Wide Missouri

JIM BRIDGER LIKED TO TELL of a place in the Rocky Mountains where rivers ran so swift that the very friction of tumbling over rocks heated the waters to boiling. No one believed him. Of course, "Old Gabe," as Bridger was known to his cohorts, had a reputation for spinning yarns and embellishing tales. But such a place did exist, and Bridger had been to it many times, even if the waters weren't *really* heated by friction. It was called Yellowstone.

Both Yellowstone and Bridger played important roles in the early history of the Rocky Mountain West. The country around the northern and central Rockies—and around Yellowstone in particular—contained a wealth of mountain streams, and over eons of time beaver found these streams to be ideal habitat. By a quirk of fashion, hats made from beaver pelts became the rage of society in the East and in Europe starting in the early 1800s. Fortunes were to be made in supplying beaver skins to these markets, and men like Bridger spent years roaming the Rocky Mountains in search of this aquatic mammal and in the process explored and mapped most of this unknown territory. These lucrative and adventurous enterprises started with one of the most incredible real estate deals in history.

When President Thomas Jefferson arranged the purchase of the Louisiana Territory from France in 1803 (for little more than the price of a modern jumbo jetliner), neither he nor anyone else knew exactly what was contained in this 828,000-square-mile surprise package. Even the precise boundaries of the territory were neither known nor stipulated, with only vague reference to the "rising of land" (presumably the Rocky Mountains) as being the western boundary. But one thing was certain: This vast domain would more than double the size of a young United States eager for expansion. And the fur traders would be the first to cash in on the abundance of new resources.

Overleaf:
Baker Lake
and Castle Peak,
White Cloud
Mountains,
Idaho

41

As early as 1540, Spaniards had explored parts of the Louisiana Territory in search of gold and the elusive Seven Cities of Cíbola, laying claim in the process to all the lands west of the Mississippi River. Spanish missions were established in northern New Mexico, and some explorers and missionaries roamed as far as central Kansas and Utah. But even after it changed hands from Spain to France, most of this region along the crest of the central and northern Rockies remained unexplored and unknown, a land of speculation and rumor and of tales brought back by wandering Indians or occasional French trappers.

Jefferson wasted no time in finding out what he had bought. Only days after the purchase, he commissioned Captains Meriwether Lewis and William Clark to explore this territory with the hope of finding an easy passage, perhaps completely by water, up the Missouri River and eventually to the Pacific shore. Though the purpose of this now famous expedition was primarily commercial, perhaps in part to justify the expenditure for the purchase, there were more subtle reasons as well: An easy passage to the Pacific Ocean would reinforce American claims to the Oregon Territory, a rather large piece of real estate in its own right, embracing most of the present-day states of Oregon, Washington, and Idaho, and a good-sized chunk of British Columbia. The Oregon Territory was also being claimed by the British, and their claim had been strengthened by the first trans-Rocky Mountain expedition to the Pacific, made in 1793 by Alexander Mackenzie. (Mackenzie's route across the Canadian Rockies proved too rugged for commercial purposes, however.) Regardless of intent, Lewis and Clark's expedition became one of the greatest in the annals of exploration.

Leaving St. Louis on the 14th of May in 1804, the Lewis and Clark party, consisting of 31 volunteers, traveled by keelboat up the Missouri River. From the beginning of their journey, they made note of the large game herds and bountiful beaver along the way as they poled and paddled up the sleek and placid river. By autumn they had reached the mouth of the Knife River in North Dakota. Here they decided to build shelters and to spend the winter with the Mandan tribe. The decision was a wise one because the winter was long and bitter, and in addition the Mandans gave Lewis and Clark much information about the land and the Indians up around the headwaters of the Missouri, information that would prove valuable when they continued the trip in the spring. But perhaps the most fortuitous circumstance of their stay was meeting and engaging the guiding and interpreting services of a trapper named Touissant Charbonneau. He would be accompanied by his 16-year-old Shoshoni wife, sold to him as a slave when she was only 11 years old. This girl, called Sacajawea, gave birth to a child that winter, then left, with the child and her husband, with the Lewis and Clark party in the spring. Though Charbonneau proved quite useless, Sacajawea's knowledge of the land where her people lived—the country along the crest of the Rockies at the headwaters of the Missouri—was of immense value, as was her kinship with the Shoshoni. As they approached the headwaters of the Missouri, the party was forced to leave their boat and proceed overland. Having made friends with the Shoshoni people through Sacajawea, Lewis and Clark were able to obtain horses, without which the success of their journey would have been doubtful.

The group crossed the Continental Divide and entered the Pacific drainage for the first time at Lemhi Pass, located on the border between what are now Idaho and Montana. They followed the Lemhi River to its confluence with the Salmon, which Clark dubbed Lewis's River in honor of his partner. Their hopes that this beautiful river would provide them easy passage to the Columbia were soon dashed upon discovering that the Salmon and its deep gorge were impassable. On the advice of their Indian guides, they headed north over the Continental Divide again and into the Bitterroot Valley of Montana. There they found a relatively easy passage over the northern Rockies at Lolo Pass, entered the Pacific drainage once more, and followed the Lochsa River to its confluence with the Snake River. For the remainder of their journey, they followed the Snake, then the Columbia, arriving on the Pacific shore on the 7th of November in 1805. Here they established camp for the winter.

In March, 1806, the party started for home, following much the same route until they reached a place they called Traveler's Rest, near present-day Missoula, Montana. Here the party divided in order to make additional explorations. Lewis's contingent headed northward to explore the Marias River before rejoining the original Missouri River route, while Clark went overland to the Yellowstone River, which he followed to its confluence with the Missouri. Lewis's group finally overtook Clark's near the mouth of the Little Missouri, and they continued down the Missouri together. When they arrived back in St. Louis in September of 1806, the expedition was hailed as a complete success. The meticulous notes and observations made by both Lewis and Clark were valuable records of plant and animal species and of the native people who inhabited these lands and were now the unwitting subjects of a new and expanding nation. Though Lewis and Clark did not find that elusive Northwest Passage, the claim to the Oregon Territory had been strengthened, and the way was now cleared for further exploration.

To do justice to the dangers and hardships of such an expedition is difficult today when the journey which took Lewis and Clark two years, four months, and ten days, may be repeated in a few days by car and with all the comforts of sanitized motels and sterilized food, the greatest danger that of a head-on collision at 60 miles per hour. In many ways the Lewis and Clark party was extremely lucky. Some encounters with the Indians might have proved fatal to some or all members of the group. (The only fatality involving an Indian was the killing of a Blackfoot brave by one of the members of the party, an act that, as we shall see later, was to cause great grief for future white men in the region.) They were fortunate also in averting starvation, for in portions of their trip game was extremely scarce, and they were forced, on occasion, to slaughter their horses or to eat dogs supplied to them by Indians. But they returned, they revealed to the world the wonders of what they had seen, and they sparked the imagination of the nation. It was, in all ways, an epic journey.

．．．．

Mammoth Hot Springs,
Yellowstone.
Photograph by
William Henry Jackson,
1871.

Sawtooth Lake
and Mt. Regan,
Sawtooth Wilderness Area,
Idaho

The Lewis and Clark expedition opened the era of commercial exploitation of the Rocky Mountain region. On their return journey, they met trappers following their original path, and one member of the expedition asked for and received permission to leave in order to search for beaver in the unknown, unexplored upper reaches of the Yellowstone River. This man's experiences would write the next chapter of exploration of the Rockies. His name was John Colter.

Colter is a mysterious figure in the history of the West, largely because he kept no journals, and his extensive wanderings and adventures have been pieced together painstakingly from other sources. Born in Virginia, he was an experienced Indian fighter in the Midwest before signing on with Lewis and Clark. The returning Lewis and Clark party encountered two trappers headed upstream on the middle section of the Missouri, and the trappers invited Colter to join them. One might assume that two years of hardship and danger were enough for any man, but Colter perhaps saw this as an opportunity to profit from his newly acquired knowledge of the country. Whatever his motives—adventure or profit— the trio disappeared back into the wilderness, and Colter was not seen again until the spring of 1807 when, on his way home again, he once more encountered fur trappers and once more was enticed to guide and trap for them.

Though Colter's first wanderings went unrecorded, there is ample evidence as to the extent of his second journey. Following the Wind River to its source, he crossed the rugged northern end of the Wind River Range at what was later known as Union Pass. He entered Jackson Hole, becoming the first white man to view the rugged and spectacular Teton Mountains, then crossed the range at Teton Pass and discovered another lovely valley, Pierre's Hole (later to become known as Teton Valley, Idaho). Here he spent the winter of 1807–08 in a camp at the base of the western slopes of the Tetons. Continuing his trek in the spring, he

Beehive group of geysers, Yellowstone. Photograph by Jack Hillers, ca. 1872.

circled around the northern and more gentle part of the Tetons, entering what is now Yellowstone National Park. His tales of geysers and boiling mud pots and smoking mountains were later ridiculed when he returned to civilization, and the existence of "Colter's Hell" was not to be confirmed until the area was rediscovered many years later. But despite the skepticism he generated on his first encounter with the wonders of Yellowstone, Colter's general information about the region around the headwaters of the Snake and Green and Wind rivers was to prove valuable for the hordes of trappers to come. And come they did. Only two years after Colter's pioneering journey, a party of trappers led by Andrew Henry built a fort on the North Fork of the Snake River not far from the western Tetons and Colter's winter encampment. Then, in 1811, a party funded by John Jacob Astor's American Fur Company trekked overland from the Missouri River to the Teton country, very nearly tracing Colter's first route. Led by Wilson Price Hunt, this group continued on, following the length of the Snake River, where they blundered into and discovered—and very nearly perished in—Hells Canyon. Hunt's party went on to the Pacific shore and established a fort—Astoria—for purposes of exploiting the rich fur trade of the Northwest.

Though others penetrated this country in the next several years, the Rocky Mountain fur trade didn't really begin to boom until the early 1820s, the beginning of the golden era of the Mountain Man. Some of the most colorful history of the Rocky Mountains began with an innocuous looking ad in 1822 in a Missouri newspaper. William Ashley and Andrew Henry, co-owners of the Rocky Mountain Fur Company, "wish to engage One Hundred Men," in the words of the advertisement, "to ascend the Missouri to the Rocky Mountains, there to be employed as Hunters." The annual compensation was to be $200, which the ad stated, and the employees would be assured of many dangers, which the ad didn't state.

In those years, St. Louis was the very edge of American civilization, the frontier town which attracted many adventurers. It was the jumping-off place for the wilderness, and Ashley and Henry soon found themselves with a number of willing and eager employees, only a few of whom were experienced at coping with a hostile wilderness environment. Nonetheless, they would soon learn or perish in the process.

Among those first employees of the Rocky Mountain Fur Company were men whose names were to become legend in the history of the West: Jedediah Smith, James Bridger, Milton and William Sublette, Thomas Fitzpatrick, Hugh Glass, and, later, David Jackson. Some of these men were as green as they come. Bridger, for example, was an apprentice blacksmith. But they learned, and learned fast.

The business of trapping was tough and dangerous, and most of it was conducted at times of year when the Rocky Mountain weather was bad. Pelts were in their prime in fall, winter, and spring; but since winter trapping was virtually impossible, the mountain men worked hardest in spring and fall. They often worked in pairs, roaming over a large expanse of country, setting traps and later returning to collect their quarry. The skins had to be scraped and dried as soon as possible to keep them in marketable condition. Wading through frigid streams and beaver ponds, a necessity if you were trapping beaver, was uncomfortable at best and at its worst led to a frequent malady of trappers—rheumatism. But even the discomforts of the profession were small in comparison to the dangers. This wild, rugged country was harsh and unforgiving to the unwary and the unprepared. Death waited at every river crossing, mountain pass, canyon entrance, or trail bend, and an unwary trapper might die at the hands of hostile Indians, drown in a swollen river, freeze to death, or even, on occasion, starve. On top of this were threats from grizzly bears that roamed over a wide extent of this country; and on at least one occasion, an outbreak of rabies in the wolf population posed serious threats when rabid wolves invaded camps.

Much of the Indian danger that the trappers faced could be traced to the Lewis and Clark expedition. During their return to civilization in 1806, a member of that party killed a Blackfoot Indian who was stealing rifles. From then on, the Blackfoot people and tribes friendly to them became mortal enemies of all white men, and they pursued their vengeance with a brutal passion. On occasion the trappers themselves were to blame for continued bad feelings between the various tribes and the white men, for they were frequently ruthless in dealing with the Indians, friendly or otherwise. A great many of these men were social mis-

Granite Creek,
Gros Ventre Mountains,
Wyoming

Aspen forest

fits and could barely get along with each other, let alone with people who had cultures and values radically different from their own. The practice of taking scalps, for example, originated with the early Spanish explorers, who used the scalps as a means of claiming bounties on certain enemies. The trappers used scalping as a grisly way to show off their skill as fighters, and more than one mountain man proudly wore his collection of "topknots" on his belt for all to see. Eventually some Indians adopted the taking of human scalps, but there is evidence that many Indian peoples regarded this perverse practice of white men to be barbaric.

Forced to adapt to this wild and hostile environment, the mountain men developed skills that seemed, at times, to surpass even those possessed by the Indians. The wilderness sustained them, but they had to learn how to harvest this bounty; or, in various places or at various times of year when it was not so bountiful, they had to learn to scavenge and scrounge for food. The wilderness also was a constant source of danger, and so they learned to recognize ominous signs, to read the land, the wind, and the waters, and to interpret these readings. The more successful mountain men—that is, the ones who survived—developed a sixth sense, a way in which they attuned their bodies and minds to this country. More than one of them claimed to have developed an acute sense of smell, the ability to smell a Blackfoot or a grizzly a mile away. And it's likely they did possess such skills. Just the intensity with which the mountain men observed their surroundings provided them with much information. The way grasses were trampled or the sudden movement of a buffalo herd might alert them to impending danger. They became fluent in the language of tracks and could tell, for example, that a party of six Blackfeet had crossed the path two days before, carrying heavy loads.

Besides these acute senses, the mountain men possessed an incredible toughness, a durability that was tested time and again. Jim Bridger carried an Indian arrowhead in his back for three years before it was removed in a crude sort of surgery by missionary Dr. Marcus Whitman on his way to the Oregon Territory. Osborne Russell and a companion were attacked by Blackfeet near Yellowstone Lake and, bristling with arrows in their legs and bodies, they walked and crawled a hundred miles, over the Teton, Big Hole, and Snake River mountain ranges to Fort Hall on the Snake River Plain. Hugh Glass was mauled horribly by a grizzly bear and left to die by his companions, but he crawled several miles to a Sioux village and eventually recovered to continue his trapping. And then there was the incredible tale of the first mountain man, John Colter himself. In the fall of the same year that he made his extraordinary journey in the Tetons and Yellowstone, Colter was captured by a party of Blackfoot Indians in Montana, just north of present-day Yellowstone Park. His partner was killed immediately in the skirmish, but Colter was to provide amusement for the Indians. Stripped naked, he was given a 30-yard head start to run, literally, for his life. He did. Streaking barefoot across sage, cactus, and rocky ground, Colter outdistanced his antagonists, paused at one point to kill an especially speedy pursuer with his own spear, and continued on to a river where he spent several hours in frigid water under a log jam while his enemies swarmed around looking for him. Then he made his way, still naked and barefoot and weaponless, a hundred miles to the nearest trappers' encampment. Not long after this, Colter decided not to push his luck any farther and he retired from the wilderness.

While there were many such tales of incredible spirit and toughness, a good many other mountain men were not so fortunate. In fact, those who survived to a ripe old age admitted that luck had played a large role in their survival. For despite all of their skills and durability, many were felled by disease and infection that might have been cured in towns and cities back east. Tetanus, smallpox, and dysentery took their toll, as did accidents and injuries. Despite the luck possessed by Russell and Bridger and a few others, many died of infections caused by Indian arrows, and a great many more were to lose arms or legs in a last-ditch effort to stem the spread of infection.

Toughness may have been one of the prime characteristics of the mountain men, but a great many of them exhibited a poetic sensitivity to the beauty of the land. Not a few of them had their favorite places. For Davey Jackson it was that lovely valley—the trappers called such valleys "holes"—on the eastern side of the rugged Tetons. Thus it became known as Jackson Hole. On the other side of the Teton Range was Pierre's Hole, named for a French trapper who passed that way around 1818 with Donald Mackenzie. (It was another Frenchman in this group, so the story goes, who named this famous mountain range. From the west the view of these sharp peaks prompted the name *Les Trois Tetons*—The Three Breasts.) Jim Bridger thought Pierre's Hole the most beautiful valley in the world. Osborne Russell, who worked for Bridger in later years, fell in love with the Lamar River country in Yellowstone and on his first visit there wrote in his journal, "For my own part I almost wished I could spend the remainder of my days in a place like this where happiness and contentment seemed to reign in wild romantic splendor surrounded by majestic battlements which seemed to support the heavens and shut out hostile intruders." On another visit a year later he wrote, "There is something in the wild romantic scenery of this valley which I cannot nor will I, attempt to describe but the impressions made upon my mind while gazing from a high eminence on the surrounding landscape one evening as the sun was gently gliding behind the western mountain and casting its gigantic shadows across the vale were such as time can never efface from my memory."

Some philosophized about their way of life, and through these reflections it is possible to gain some insight into their lives and times. Warren Angus Ferris arrived on the scene in 1830 in the employ of John Jacob Astor's American Fur Company, known simply as "The Company" to all of the trappers. Of his life here for the next several years, Ferris wrote:

> Strange, that people can find so strong and fascinating a charm in this rude and nomadic, and hazardous mode of life, as to estrange themselves from home, country, friends, and all the comforts, elegances, and privileges of civilization; but so it is, the toil, the danger, the loneliness, the deprivation of this condition of being, fraught with all its disadvantages, and replete with peril, is, they think, more than compensated by the lawless freedom, and the stirring excitement, incident to their situation and pursuits. The very danger has its attraction, and the courage and cunning, and skill, and watchfulness made necessary by the difficulties they have to overcome, the privations they are forced to contend with, and the perils against which they must guard, become at once their pride and boast. A strange, wild, terrible, romantic, hard, and exciting life they lead, with alternate plenty and starvation, activity and repose, safety and alarm, and all the other adjuncts that belong to so vagrant a condition in a harsh, barren, untamed, and fearful region of desert, plain, and mountain. Yet so attached to it do they become that few ever leave it, and they deem themselves, nay are, with all these bars against them, far happier than the in-dwellers of towns and cities, with all the gay and giddy whirl of fashion's mad delusions in their train.

Were their lives enriched by these experiences? According to Osborne Russell they were. Encamped for the winter near his beloved Lamar River in Yellowstone, he wrote:

> We all had snug lodges made of dressed Buffalo skins in the center of which we built a fire and generally comprised about six men to the lodge. The long winter evenings were passed away by collecting in some of the most spacious lodges and entering into debates, arguments or spinning long yarns until midnight in perfect good humour and I for one will cheerfully confess that I have derived no little benefit from the frequent arguments and debates held in what we termed The Rocky Mountain College and I doubt not but some of my comrades who considered themselves Classical Scholars have had some little added to their wisdom in these assemblies however rude they might appear.

Classical scholars? The vast majority of the mountain men could neither read nor write. Nonetheless, the lure of adventure did attract men of widely varying backgrounds, some of them highly educated. It was, as DeVoto so well pointed out, a time of fulfillment.

Bear Creek, Idaho Primitive Area, Idaho

Bighorn sheep along Salmon River, Idaho

The geographic center for the fur trade was located at the headwaters of the Green, the Snake, and the Yellowstone rivers amid country that would cause most men to wax poetic. The mountains themselves were awe-inspiring: the Absarokas, the Beartooth, the Wind Rivers, the Gros Ventres, and, of course, the magnificent Tetons. In addition to the major rivers, there were numerous smaller ones, each beautiful: the Hoback, Ham's Fork, the Bear River, the Wind River, Jefferson's Fork, Henry's Fork, the Teton River, plus a host of smaller streams and tributaries, few of which remained unknown to the far-ranging mountain men.

It was by no accident that the mountain men focused their attention on this geographic area. Their stock-in-trade—beaver—were plentiful in this country. South of here, on into Utah and western Colorado, the land became hotter and drier, less conducive habitat for the aquatic mammals. To the north, in central and northern Montana, the danger posed by the Blackfoot people increased. (It was perilous enough even here.) But perhaps the most important factor was the discovery of an easier access to the upper Green River country:

Tower Falls,
Yellowstone.
Photograph by
William Henry Jackson,
1872.

South Pass. Located just off the southern end of the Wind River Range, it is not really a pass at all, but a high, flat prairie separating the Wyoming Rockies from the Colorado Rockies. The route to beaver country followed the North Platte River upstream to the Sweetwater River, then went along the Sweetwater to South Pass. Once across the pass, it traversed a short stretch of desert to the Green River, which was followed to its head-waters. Those who wished could continue northward to the Snake and Yellowstone country. In later years, the route across South Pass was to provide the easiest access to the West for the migration to Oregon. For the mountain men this route facilitated the passage of pack trains and later the wagons that brought them supplies and provided a means of transport-ing their wares back to market. The discovery of the route also gave impetus to the estab-lishment of one of the more famous institutions of the fur trade: the Rendezvous.

Starting in 1825 and continuing through 1839, the Rendezvous became an annual month-long affair that was payday, trading mart, celebration, and three-ring circus all rolled into one. At times it had all the earmarks of a full-scale riot. When everyone met at the agreed-upon site for Rendezvous—and it varied from year to year—the trappers sold the skins collected over the past 11 months and if they'd had a good year were suddenly en-riched by one to two thousand dollars. The trader, in addition to being purchasing agent for a fur company, was also a mobile company store. He packed in (or in later years brought in by wagon) whiskey, tobacco, powder, bullets, flour, salt, coffee, plus the beads and bau-bles that the mountain men used for Indian trading or buying an Indian wife. All these items were, of course, sold at a substantial profit to the trader (sometimes at an estimated 2,000 percent markup over St. Louis prices), and since there was no competitor to bargain with, the mountain men had to pay the exorbitant prices. Many of them didn't seem to mind as long as there was plenty of whiskey.

In addition to the trappers, friendly Indian tribes joined in the celebration. Many of the Indians learned that there was a handsome market in women, and thus wives were bought and sold; or merely the temporary favors of an Indian lass were paid for by some beads, bullets, or powder. In the case of an especially desirable woman, a rifle or horse might be offered. For some 30 days during the Rendezvous, the mountain men whored, drank, gam-bled, drank, brawled, and drank. Sometimes the entertainment got a little rough, as when old Joe Meek herded an angry grizzly through camp as a joke, scattering people in every direction. And at the Pierre's Hole Rendezvous in 1832, one old trapper, passed out temporarily from booze, was doused in alcohol and ignited by his drunken friends. (He survived, apparently the only damage being a few burnt whiskers.) Kit Carson, who was to go on to even greater fame as army scout and Indian butcher, took on a hulking, threatening drunk of a trapper and shot him dead at one of the Green River Rendezvous. And then there was the time when . . . Well, it's a long list, and in the telling and retelling many incidents grew to legend pro-portions, so that it's difficult to separate fact from fantasy. But one thing is certain: The mountain men lived a life few can imagine today, and if some of them behaved as though each day were their last, they must be forgiven. Frequently it was.

· · · ·

The decline started in the mid-1830s. Another whim of fashion had caused the de-mand for beaver to decline sharply. By 1840 or so, it was over. Slowly the fur companies dissolved or went broke or both. And the mountain men began drifting away. Earlier, such men as Jed Smith and Joe Walker had pioneered routes to California, so some of them headed for that promised land. Others moved on to the Southwest, among them Kit Carson, who became guide, scout, and hero of dime novels back east. Jim Bridger stayed on for a while, guiding the trickle of migrants who were on their way to another promised land, Oregon. Not far from South Pass, he established a trading post, and for several years Fort Bridger was an important stop on the way West. A great many of the mountain men faded into obscurity, but there is no doubt that they all looked back upon their experiences in the Rocky Mountain West with pride and perhaps some nostalgia. Few other eras of American history packed so much adventure into such a short period of time.

Upper and Lower
Mesa Falls,
Henry's Fork of
the Snake River,
Idaho

Paths Across
the Continent

I'm not sure what I expected to find here. I mean, surely the trail has been picked clean of all relics in more than a century of time. There are no broken wagon wheels, grease buckets, water barrels, or cast-off possessions. And no blanched and bleaching bones of poor souls, human or animal, who didn't make it. In fact, as I stand here in the middle of these two ruts, a Coors beer can glinting at me in the sunlight and the rumble of a jet overhead somewhere, I wonder if this is even a part of the original Oregon Trail, for this section of the Red Desert has been crisscrossed in many places by jeep roads in recent years, most of them probably from oil exploration crews. In fact, where I left the interstate some 50 miles back, I could see several oil pumps, their great hammer-shaped flywheels rotating in that slow motion way that they have while sucking the black fluid from deep in the ground. No, I really can't be sure that this is the trail, but then I suppose it doesn't matter. It's the feeling of this place that I'm after, the sense of fear, apprehension, sorrow, joy, hope. But I'm not sure I'll find it.

The wind is blowing steadily. The wind always blows steadily across southern Wyoming, prompting numerous jokes. Question: Does the wind ever stop blowing here? Answer: I don't know, I've only lived here 15 years. Today it is a hot, dry, lip-parching wind that offers no cooling relief from the blazing sun. In winter it is a bitter cold wind that knifes through you. "The snow doesn't melt here in these parts," goes yet another story. "It just blows around 'til it wears out." The tracks lead west across rolling, sage-covered land toward South Pass, some 30 miles or more away. It is a land not without a certain amount of beauty. But then I've always been a bit of a desert rat. I like the simplicity and expansiveness of it all. I suppose, however, that I might view it a little differently if I were traveling across it at a snail's pace in an oxcart more than a century ago.

THE MIGRATION BEGAN EVEN BEFORE the mountain men had moved on and it followed the paths established by them only a few years before. Among the first of the great immigrant parties was that of Dr. Marcus Whitman, Methodist minister and missionary. In 1843 he led a group of almost 500 people to the promised land of Oregon and the mission he had established there. From Missouri they followed the North Platte River across what are now Nebraska and Wyoming, passed north of the Red Desert along the Sweetwater River, and headed over South Pass, pressing on to Fort Hall and then across the brutal Snake River Plain to the green and fertile valleys west of the Cascade Mountains. Their track became known as the Oregon Trail, and the next few years saw thousands repeat the journey.

What forces motivate men to toss off security, pull up stakes, and move on? The dream of wealth? The hope for land? The spirit of adventure? The search for Nirvana, contentment, peace? Or dissatisfaction, greed, restlessness? Yes, all of these probably, and perhaps a great many more subtle and inexplicable factors as well.

To understand more fully those motivating forces one must also understand the times. The little nation of America was flexing its expansionist muscle, fueled and fired by the Louisiana Purchase and by the growing tales of that fabulous and frightful land. But more important, farther west were Oregon and California, and certain politicians were eyeing those territories as potential additions to the nation. Where America's frontier then stood, in the Great Plains, the land was harsh. Fertile enough, perhaps, but not friendly. Cold winds and snow raked the land in winter, and the sun beat down unmercifully on the unshaded prairie in summer. Ah, but Oregon and California were gentle and sweet, fertile and fecund. There a man might never feel the sting of winter again, nor toil so hard to make things grow. Thus a lustful eye was cast on these lands by the Pacific shore.

But getting to them was another matter. Even before the return of Lewis and Clark, Zebulon Pike had set out to explore the central Rockies, and his report was not at all optimistic. "The Great American Desert," it came to be called, a label applied to all the Rocky Mountain West; and to reach the fertile lands farther west, it had to be crossed. But despite—or perhaps because of—these disparaging reports, there were people willing to risk the journey. Urging them on was the philosophy of the time: Manifest Destiny, the divine predestination that America should rule the continent from ocean to ocean, regardless of who happened to own the land at the time. With this the rationale, Texas and California were wrested from Mexico, the Oregon Territory liberated from Great Britain. But the ultimate incentive was land—640 acres in Oregon, for example, promised to each head of a family who would settle it.

What does 640 acres mean to a man? It means freedom and independence, those great and theoretical American ideals. Many of the people were a generation or less removed from small subsistence plots in Europe, and 640 acres—one square mile—was an empire! A man could live off the land and never be beholden again to another man. If one had to risk death to reach those lands, well, so be it. The rewards were worth it.

Somewhat later, there was gold. Instant wealth. It almost seemed to reinforce the argument of Manifest Destiny, for surely it was a sign of divine intervention. So the torrent became a flood. Twenty thousand people in 1849 alone trekked across the Oregon Trail as far as Soda Springs or Raft River, then took the California Cutoff across northern Nevada, down the Humboldt River, and over the Sierra Nevada.

In those first years of the Great Immigration, some of the mountain men found their services in demand as guides. Undoubtedly it came as a shock to them to find so many people crossing the wilderness. As the fur trade declined, some of the trappers eventually moved on with the immigrants. Osborne Russell became a prominent political figure in Oregon in later years. Still others went on to California. In time the Oregon Trail became so well traveled that the mountain men were no longer needed as guides. But the trip remained as arduous as ever, so much so that even Horace Greeley described it as "an aspect of insanity."

What was life like on the Oregon Trail? Invariably the trip began with great hope and optimism. The jumping-off places were towns like Independence, St. Louis, St. Joseph, and Liberty, Missouri. There the hopefuls gathered and organized themselves into groups and trains of wagons, purchasing goods and supplies for the long trip. In the early years some leaders barely knew the route; their knowledge of the land ahead was vague at best and totally incorrect at worst. In time, seasoned experts took command to lead the parties through the dangers of Indian and drought.

It took about 60 days to travel across the plains of Kansas and Nebraska, up the North Platte to South Pass. Yet here the journey was barely half over, and by now much of the initial joy and hope was erased by fatigue and perhaps doubt. It had been a hard journey to this point, fighting heat and dust and thirst, or rain and cold and mud, depending on the season and the quirks of weather. For women and children it was especially

Eagles Rest Peak
and Leigh Lake,
Grand Teton
National Park,
Wyoming

Oregon Trail,
Devil's Gate on the
Sweetwater River.
Photograph by
William Henry Jackson,
ca. 1875.

hard, and more than a few died on the long trip. "A great many of us died young, perhaps before our time," wrote one traveler. "It happened along the trail at Independence Crossing, Ash Hollow, at South Pass and Bear River." The river crossings were especially hazardous at times, even for the young and strong. "We soon came to the crooked, treacherous Snake River where we lost two of our men, Ayres and Stringer," wrote another immigrant, this one in 1845. "Ayres, who was an old man about sixty, got into trouble with his mules in crossing the stream. Stringer, who was about thirty years of age went to his relief, and both of them were drowned in sight of their women folks whom they had ferried across. The bodies were never recovered."

The Snake River Plain, from Fort Hall to Fort Boise, seemed endless to many of them, described by one traveler in 1844 as "the most Barren Sterril region we have yet passed nothing to disturb the monotony of the Eternal Sage plain which is covered with broken cynders much resembling Junks of pot mettal & Now & then a cliff of Black burned rock which looks like Distruction brooding over dispair." And still, from Fort Boise they had a long way to go, across the Snake River again at Farewell Bend, over the Blue Mountains to the Columbia, then down this great river to their final destination—the lush lands of the Willamette Valley.

Was it worth it? Was this Willamette Valley indeed the land of milk and honey? Not all thought so. One woman wrote in 1848:

> Rain all day. If I could tell you how we suffer you would not believe it. Our house, or rather shed joined to a house, leaks all over. The roof descends in such a manner that the rain runs right down into the fire. I have dipped as much as six pails of water off our dirt hearth in one night. Here I sit up night after night with my poor sick husband, all alone, and expecting him every day to die. . . . I have not undressed to lie down for six weeks. Besides our sickness I had a cross little babe to care of. Indeed, I cannot tell you half.

The next day her husband died. Many underwent similar hardship, some to no avail, for not all of them received the 640 acres they had been promised. But for the most part the newcomers were happy, and the flow across the Oregon Trail continued. Then, in 1849, the character of the immigration changed. The Gold Rush had started. Every get-rich-quick artist, every adventurer, every dreamer of wealth, gambler, con man, prostitute, and entrepreneur saw the fabulous gold strikes in California as a means to become wealthy beyond his or her wildest dreams. Some were unwilling to submit to the trials of overland travel, and these sailed by ship down the East Coast and around Cape Horn to San Francisco. But others took the land route, and by the tens of thousands they streamed across that beaten path up the North Platte and across South Pass, then branched off on what was now the trail to the gold fields. In the summer of 1850 alone, an estimated 55,000 travelers stopped at Fort Laramie on the Platte. Oregon immigrants continued to pour across the continent as well, though not in such great numbers as the gold seekers.

Along the
Oregon Trail, near
the Sweetwater River.
Photograph by
William Henry Jackson,
ca. 1875.

65

Near headwaters of West Dolores River, Colorado

Aspen

The '49ers, they were called, but the rush continued on into the 1850s. Those who flocked through the Rocky Mountain region considered it a land merely to be tolerated, a hardship to endure until they could reach those richer lands. There was nothing to coax them into staying. No gold. And certainly no green and fertile valleys. This was a harsh and hostile place, offering few rewards compared to California and Oregon. Then, in 1859 on Clear Creek in the Front Range of Colorado, gold was discovered, and almost overnight the tide rebounded from California, and now the rush was to Colorado. The rivers of the Rockies that had served merely as paths across the wilderness now became objects of commerce and avenues to potential wealth.

It looks like a scene right out of a World War II movie. The hills and mountainsides that roll away from my vantage point could be an Italian countryside, the scars and holes and piles of debris the aftermath of a bombing raid. Rusting hulks of buildings lie shattered and scattered all about me. What I'm viewing is not the result of a war of man against man, however, but of man against the earth. And the residue of that battle is everywhere in evidence; gleaming mounds of yellow-white tailings stand out boldly against the forest.

I'm standing on a hillside that commands a sweeping view of the country around Central City near the North Fork of Clear Creek in Colorado's Front Range. The richest square mile on earth, it was called in its day, and I have no reason to dispute that claim. Today the mines are closed (though a rumor or two says that with today's gold prices what they are, and new technology being what it is, some may be reopened). Central City and Black Hawk are now tourist attractions, "restored" to recapture a flavor of life that probably never really existed here to begin with. There is a carnival-like atmosphere in the streets of Central City, where visitors from New York and Georgia and Minnesota and California mill around on the sidewalks, poking into quaint little shops, buying postcards and curios to send or take back home. A motorcycle gang, their choppers lined up neatly next to a curb, lounge insolently and arrogantly against the walls of the old opera house. To complete the illusion of a circus, the smells of popcorn and cotton candy pervade and per-

Placer mining.
Photograph by
William Henry Jackson,
ca. 1885.

Library, The State Historical Society of Colorado

Cherry Creek,
floods in Denver,
1864.
Photographer
unknown.

fume the air. Is it possible, I wonder, to capture the spirit of this town when it first boomed? In my cynicism toward what is going on around me, I decide that it is not. Then it occurs to me that what happened here more than a hundred years ago was so amazing, so surreal, so incredibly bizarre, that maybe, in a strange sort of way, this contemporary madness does approximate it.

Strangely enough, it was merely the *rumor* of gold that touched off the rush of 1859. Heading back to his native Georgia after a summer of fruitless gold panning along the South Platte River, William Russell was plagued by the story that he had made a gold discovery in the Rockies. The more he denied it, the more he convinced people that it was true, and newspapers spread the story. Soon a small-scale migration was underway to a place at the foot of the Rockies where Cherry Creek joined the South Platte River, a spot that would come to be known as Denver in a few years. Beginning as early as December, 1858, cabins and tents and tipis were flung up along the shores of these two streams, and there was an air of expectancy as more and more people continued to arrive. Everyone was waiting for spring and melting snows so that they could head up into those mountains and find their fortunes.

In retrospect, it had the air of a self-fulfilling prophecy, almost as though the gathering hordes had willed the gold to be there. And it was. At first it was placer gold along Clear Creek, especially the northern fork of that stream where it descends through steep canyons from the high country. But placer gold does not last, and a few hundred men working a stream could deplete it in a short time. In May of 1859 lode gold, the stuff from which real fortunes are made, was found in the upper drainage of the North Fork of Clear Creek. The migration began in earnest.

West Dolores River
and Dunton, Colorado

Central City,
Colorado,
in the 1860s.
Photographer
unknown.

Even by today's standards based on high-speed transportation, the swiftness with which this area swelled in population is truly astounding. Perhaps frightening is a better word. In a space of time that could best be expressed in days rather than weeks or months, 15,000 people swarmed into the Clear Creek Basin! The incipient town of Denver was virtually emptied of all its transient inhabitants, and new arrivals didn't even bother to stop there. The throng flocked to Central City, Black Hawk, Nevadaville, Missouri City, and other assemblages of shacks and tents that had sprung up in the basin. That summer of 1859 was a frenzy of activity that is unmatched elsewhere in the Rockies. The ore was dug, towns established, services offered, goods bought and sold, people enriched, impoverished, robbed, swindled, shot, or hung. And most of it took place in the space of six months. By winter of that year the surface gold was gone, and all but a few hundred people headed back to milder Denver. In the summer of 1860, a mere thousand people resided in the Central City area. It was the few deep mines that survived and thrived. The boom had settled into a more orderly and dignified rape of the land, and although the population grew, it grew more slowly than it had in that previous frantic summer. But the rush to the Rockies was now underway in dead earnest.

Within a few years, the Central City phenomenon was repeated in numerous places along the Rocky Mountain chain. Up and down the length of the Rockies, from the Sangre de Cristos and the San Juans to the Owyhees and the Sawtooth and the Clearwater country, men with picks and pans looked for signs of the yellow metal, stopping to pan every gravel bed in every river and stream. Occasional flecks of gold in their pans spurred them on, and they followed and panned those streams to their sources in hopes of finding the fabulous

Mother Lode in the solid bedrock. When lode gold was found, it precipitated a rush to the area, and what followed was largely the same whether it occurred in Leesburg in Idaho Territory or in Telluride in Colorado. The initial discovery brought a huge influx of people, sometimes 10,000 or more; towns were established, some of them no more than a scruffy collection of shacks and shanties and tents; then, of course, everyone scrambled for the easy winnings of gold on the surface, blasting away at any promising-looking outcrops or veins, then washing the pulverized ore with sluice boxes. But in time the surface gold played out and the boom was over. The impatient and the unlucky moved on while the few who remained behind worked the slower-paying deep mines. Invariably the ones who became wealthy were not the Johnny-come-latelies who flitted from strike to strike, but the solid men who stayed put, sunk capital in elaborate and expensive deep mining operations, started banks, and opened stores to serve the now stabilized population of the small but growing mining towns. In time, of course, even the deep mines were exhausted of the glittering resource, but some of them operated into the 20th century before folding.

It was an era of amazingly swift change for the Rockies. The crazy jigsaw pattern of territories seemed to change overnight. In 1863 the Idaho Territory was established, a vast domain larger than Texas and including the modern-day states of Montana and Wyoming. By 1864, however, Montana Territory was carved out of that, and in 1868 Wyoming Ter-

Clear Creek Canyon,
near Central City,
Colorado.
Photograph by
William Henry Jackson,
ca. 1875.

Harebell

ritory was established as a separate entity. California was granted statehood in 1850, Oregon in 1859. But for those wild and woolly political domains that lay between Missouri and the Pacific shore, it was to be a decade or two before statehood would be achieved: 1876 for Colorado, 1889 for Montana, 1890 for Idaho.

The establishment of political boundaries has a profound effect on a land. It is the process by which the untamed wilderness is legitimized, formalized, and brought within the sphere of political and social control. And the process brings change, for the land and for the people.

In the decades that followed the mad lust for gold, people began to discover that parts of the Rockies were not as hostile as once believed. As early as 1847 the Mormons had settled the lovely valley of the Great Salt Lake, and by the gold rush of 1849 the population of Utah Territory had swelled to 11,000. The Mormons were not gold seekers. Escaping religious persecution back east, they sought peace and quiet and an agricultural way of life. Undoubtedly their success at making the land bloom spurred others to try. Those who followed the prospectors divided the land into ranches and farms, raised cattle or sheep, and joined in the enterprise of taming the Rocky Mountain wilderness. It did not yield easily, and still hasn't in places. But perhaps most unyielding were the native people, the Indians, who now saw their lands being overrun by madmen in search of a useless yellow metal; by fools who dug up the camas prairies to plant exotic crops; by ruthless men who slaughtered the great buffalo herds; by a people who would even change the great rivers and who thought that they *owned* their various pieces of Mother Earth.

Of all the places I've visited trying to achieve a sense of historical perspective, not many have fulfilled my expectations. Oil wells, jet contrails, and beer cans destroyed my illusion of the Oregon Trail. The gaudy tinsel and milling mobs of modern-day Central City didn't give me any true sense of that place either, any more than the swarm of aluminum campers in Jackson Hole brought me closer to the mountain men. Where the Jefferson and Madison and Gallatin rivers meet to start the great Missouri River and Lewis and Clark camped to contemplate their journey, there are now power lines, railroad tracks, a cement factory, and another of those ubiquitous interstate highways.

Some places have come close, however. Retracing a part of Powell's epic journey on the Colorado River in Grand Canyon, I was able to experience the awe and apprehension of that first party. And standing on the windy summit of Table Mountain in the western slopes of the Tetons, I sensed the excitement of William Henry Jackson loading his bulky wet-plate camera for the first photograph of the fabulous Grand Teton in 1872. But perhaps my strongest feelings have been generated at this place where I now stand on the shore of the Snake River in Hells Canyon. It is the spot where Chief Joseph led his people across this deep and raging river, seeking escape from an oppressive United States government.

This is not the deepest part of this gorge, which reaches a depth of 8,000 feet many miles upstream. Here the canyon is a mere 4,000 to 5,000 feet deep, though the total scope is not visible from these deep confines. Despite the ruggedness of the terrain, there is a softness and gentleness to it. The dark undertone of lava is covered by lush, grassy hillsides that gleam wetly in the gentle rain that is falling. The verdant slopes rise steeply away from the river toward a rim of rock and deep green forest above.

Perhaps it is the gloom of weather, but I sense the sadness of Joseph and his people at having to leave this beloved land of theirs. From the Wallowa Valley west of here, they began the long journey that would end in tragedy for them. What makes a people flee from a land that has been their home for countless generations?

The traditional home of the Nez Percé people was that area of eastern Oregon and Washington and northwestern Idaho encompassing the Wallowa Mountains and Valley, extending roughly to the Clearwater River country around its confluence with the Snake River. It was a fertile land, providing the people with abundant food. The camas root was plentiful, and game abounded in the rolling hills and gentle valleys. Then there were the great rivers: the Grande Ronde, the Imnaha, the Clearwater, and, of course, the Snake, sometimes called the Holy Mother of Waters. These were a bountiful resource in themselves with the great salmon migrations each year.

It was an expansive and varied land, with rugged mountains, awesome gorges, and beautiful rivers, the kind of land that could provide the people with nourishment for both body and soul. Perhaps the nature of a land shapes the nature of a people, for the Nez Percé had a peacefulness and gentleness to match their country. Lewis and Clark found them a warm and friendly people who were only too happy to aid them in their journey. Even with the influx of settlers to the Oregon Territory, which must have had ominous overtones to the Nez Percé, their friendliness never diminished. And later there emerged yet another virtue: patience. Patience with all of the vagaries and treacheries of those who were invading and appropriating their homelands.

In 1855 at a council meeting in Walla Walla, many Nez Percé leaders were persuaded by the U.S. government to sign a treaty which delineated the lands to be reserved to the Indians and those to be open to white settlement. But Old Chief Joseph refused to sign, saying that a man could not sell what he did not own, and that no man owned any part of Mother Earth. He took his people away from the council meeting, back to their lovely Wallowa Valley.

The treaty was broken almost as soon as it was signed, for the government would not—or perhaps could not—stop settlement and appropriation of these lands by the immigrants. But the worst blow was the discovery of gold in the early 1860s in the Clearwater country. Immediately the Nez Percé lands were overrun by hordes of money-mad men who neither knew nor cared about any treaties.

For the Nez Percé, as for Indians elsewhere in the West, it was difficult to understand the strange and distorted values of these white invaders. The native people lived in harmony with their land, harvesting such native plants as the camas root, hunting the game, and fishing the streams for their food. The land was something treated with reverence, for it sustained them. Mother Earth, it was frequently referred to, and ownership of the land was a concept foreign to them, as was drastic alteration of that land to grow crops or raise cattle, or worse, to blast and dig for metals of dubious value. The intruders, on the other hand, were people caught up in the great American expansionist dream, who saw this a rich and fertile place, free for the taking, and inhabited only by a bunch of lazy, primitive —and occasionally bloodthirsty—savages who were beyond hope of salvation. In these trying times, it was Smohalla, Nez Percé philosopher, who best expressed the disparity of cultural values:

> My young men shall never work. Men who work cannot dream; and wisdom comes to us in dreams. You ask me to plow the ground. Shall I take a knife and tear my mother's breast? Then when I die she will not take me to her bosom to rest. You ask me to dig for stone. Shall I dig under her skin for her bones? Then when I die I cannot enter her body to be born again. You ask me to cut grass and make hay and sell it and be rich like white men. But how dare I cut off my mother's hair?

For almost two decades the Nez Percé negotiated, argued, discussed, and renegotiated their treaty with the U.S. government. Each time, the treaty was ignored or broken as more and more Nez Percé lands were appropriated and settled. Old Chief Joseph died in 1872, but before his death his son, the new chief, promised never to yield the land of their beloved Wallowa Valley. It was a promise that Young Joseph found increasingly difficult to keep.

The final blow came in 1877 with the conclusion on the part of the government that it was the Nez Percé people who were the cause of all the problems. The solution was to move them to the confines of a small reservation on Lapwai Creek near the Clearwater River, a site far too small to sustain them in the way of their expansive Wallowa Valley. General Oliver Otis Howard delivered the ultimatum: Move or be moved forcibly.

On the night before the deadline, a party of young militant Nez Percé warriors staged a raid on a small white settlement near the Salmon River, nearly wiping out the entire pop-

Beaver dam

ulation. This sealed the fate of the Nez Percé people. When Joseph learned of the killings, he realized that reprisals were inevitable and that to stay and fight would be futile. Thus, in early June of 1877, he began a long retreat across the northern Rockies to seek refuge in Canada. More than 600 of them, men, women, and children, began the journey, crossing the Wallowa Mountains, down into and across Hells Canyon and the Snake River, headed toward the Montana border. In White Bird Canyon near the Salmon River, Joseph and his warriors scored their first military victory when they nearly annihilated a detachment of troops. Then, for nearly a thousand miles, up the Clearwater, over Lolo Pass, down the Bitterroot Valley and into the Big Hole Valley of Montana, through Yellowstone, and then north, Joseph and his men outwitted and outfought the combined forces of several armies of the U.S. government. It is probably the ultimate irony that their path nearly retraced in reverse the route used by Lewis and Clark, the first white men to encounter the Nez Percé some 70 years earlier.

In time their skill and luck gave out. Weakened by several encounters with troops, slowed by the aged and infirm and the very young, the Nez Percé were finally caught a mere 40 miles from their goal, the Canadian border, in the snowy Bear Paw Mountains of northern Montana. It was here that Chief Joseph delivered his famous surrender speech:

> Tell General Howard I know his heart. What he told me before I have in my heart. I am tired of fighting. Our chiefs are killed. . . . The old men are all dead. It is the young men who say yes or no. He who led on the young men is dead. It is cold and we have no blankets. The little children are freezing to death. My people, some of them have run away to the hills and have no blankets, no food; no one knows where they are—perhaps freezing to death. I want to have time to look for my children and see how many of them I can find. Maybe I shall find them among the dead. Hear me my chiefs. I am tired; my heart is sick and sad. From where the sun now stands I will fight no more forever.

Despite assurances that they would be sent to the Lapwai Reservation in Idaho's Clearwater country, the tribe was taken to Fort Leavenworth, Kansas, and from there, in 1878, to Indian Territory (Oklahoma). Though bent, Joseph's spirit was never broken, and he pleaded for several years that his people be allowed to return to their native lands. "You might as well expect the rivers to run backward," he argued, "as that any man who was born free should be contented penned up and denied liberty to go where he pleases." So many of the Nez Percé sickened and died in the barren plains of Indian Territory that in 1885 the government relented, and a small group of women and children were allowed to return to the Clearwater country. Joseph never again saw even that part of his native land. Along with others considered "dangerous," he was sent to the Colville Reservation in northeastern Washington, where he was still confined at the time of his death in 1904.

> The white men were many and we could not hold our own with them. We were like deer. They were like grizzly bear. We had a small country. Their country was large. We were contented to let things remain as the Great Spirit made them. They were not, and would change the rivers and mountains if they did not suit them.
>
> Chief Joseph

Journey into the Great Unknown

ON THE 24TH OF MAY in 1869 began one of the great river journeys in history. It started inauspiciously from the little settlement of Green River, Wyoming, then known as Green River Crossing. There, as described by the leader of the expedition, "the good people of Green River City turn out to see us start. We raise our little flag, push the boats from shore, and the swift current carries us down." "Down," for that party of ten men in four boats, was to be a wild, and at times, a frightening trip of more than 1,000 miles on first the Green and then the Colorado rivers. They would not be seen again until nearly September of that same year, and before their return to civilization, reports would circulate widely that they had perished on their daring adventure.

The leader of that expedition was Major John Wesley Powell, a teacher, geologist, and one-armed veteran of the Civil War. As a youngster, Powell had taken a keen interest in natural history and geology and had a lust for adventure, wandering on foot and often alone through many still-wild areas of the East and Midwest. In 1867, as a professor of geology at Illinois State Normal University, he led a natural history expedition to the Colorado Rockies, returning to those mountains again in 1868. During the winter of 1868–69 he explored parts of the Green River country where he became intrigued—some say obsessed —by the vast, unexplored country of the Green and middle Colorado rivers. And by May of 1869 he was back with a modestly funded expedition to explore the unknown Colorado River and its tributaries.

In 1869 there were few large areas of the American West that remained unexplored. What the fur trappers hadn't roamed, the gold seekers and the settlers did. California and Oregon had already attained statehood and were on their way to becoming tamed and cultivated. Boom towns had sprung up on the heels of gold discoveries all across the Rockies. On the South Platte River, a little settlement called Denver was a thriving and growing metropolis. And the new transcontinental railroad had already pushed through such frontier towns as Cheyenne and Green River and Salt Lake City. At this time only two sizable regions of the West could truly be called *terra incognita*. One was located around the headwaters of the Yellowstone River in northwestern Wyoming, a place purported to contain smoking mountains and heated rivers, the existence of which had yet to be confirmed. The other region was the middle part of the drainages of the Green and Colorado rivers.

The Green River, which should rightfully be called the headwaters of the Colorado, begins its life as a brawling mountain stream tumbling out of the northern Wind River Range of west-central Wyoming. Well known to the mountain men, the upper Green winds southward through open and gentle country. Near the present-day border between

Salmon River,
Idaho Primitive Area,
Idaho

Lupine, Indian paintbrush, and balsamroot

Wyoming and Utah, the character of the country suddenly changes, and the Green enters the desert and sandstone topography that will mark the remainder of its journey. The Colorado begins in much the same way as the Green, draining most of the central Rockies with such tributaries as the Gunnison and the Dolores. By the time the Green joins forces with the Colorado in eastern Utah, they are both sizable rivers, turbulent and swollen from a large part of the Rockies' runoff. And it's with the great fluid momentum transmitted by the Rockies that the Colorado system has created a network of spectacular and rugged gorges, carved across hundreds of miles of colorful sandstone country in Utah and Arizona. It's known variously as Canyonlands, the Plateau Province, or simply Slickrock Country, described by Powell himself as "a land of desolation, dedicated forever to the geologist and the artist, where civilization can find no resting place." Present-day author Edward Abbey calls it "the least inhabited, least inhibited, least developed, least improved, least civilized, least governed, least priest-ridden, most arid, most hostile, most lonesome, most grim bleak barren desolate and savage quarter of the state of Utah—the best part by far."

It is almost impossible to describe to someone who has never seen it, for it is a region like few others on earth. Visualize a plateau beginning in the northeast corner of Utah and western Colorado, a desert and sandstone plain sloping south and westward away from the outwash of the Rockies toward Arizona and the Gulf of California beyond. Then picture the deep gorges of the Green and Colorado rivers, steep, sheer-walled, and up to 3,000 feet deep in places, which join to form a "Y" whose tail, the Colorado, continues on across that plain. Next you must picture deeply incised tributaries—and there are many, all with steep-walled canyons—that come down to join the Colorado and further carve that plateau into an infinite network of isolated mesas and buttes. Then imagine the endless tributaries to the tributaries also etched into that slickrock country and you may have a rough mental portrait of a crazy land of red walls, isolated fins and buttes, eroded and corroded towers, pinnacles, turrets, stone goblins, and haunting sculpture; a region that is real and surreal, bleak, barren, and beautiful all at once. And finally, as though all this were not enough to saturate the senses and boggle the mind, cap it all with the scenic climax, the wonder of wonders, the vast, awesome, magnificent Grand Canyon; then you may have some faint notion of what this incredible land is all about. Maybe.

And yet, for all its forbidding nature, this land has not been completely devoid of man's touch. For centuries, the Anasazi and other canyon and desert cultures flourished here. But around A.D. 1300, they moved on, and by the time the first white explorers reached this country, it was largely devoid of people. The hot desert winds whistled through silent canyons and through the ruins of the civilizations that had survived here.

The first white men were Spaniards. Francisco Vásquez de Coronado reached western New Mexico in his expedition of 1540 and there he dispatched a party to look for a great river described to him by the Indians. This little group, led by Captain García Lopez de Cárdenas, made it to the south rim of the Grand Canyon, but the honor of being the first white man to see this wonder left Cárdenas unimpressed. Undoubtedly disappointed at finding this terrain so rugged and impassable, he reported back to Coronado that it was a "useless piece of country."

For more than two centuries after that, Spanish trappers, explorers, and missionaries roamed the fringes of the canyon country, but the heart of this wilderness remained largely unexplored and unmapped. In 1776 two priests, Francisco Atanasio Domínguez and Francisco Silvestre Velez de Escalante, left Santa Fe on a trek to find a route to Monterey, California, where other Spanish missions and settlements were located. Although the trip was a failure (they never even got close to Monterey), it was significant in that they made a grand loop completely around this unknown canyon country, heading north through Colorado, discovering the Green River, then going west to Utah Lake and south along what is now the Utah-Nevada border. Near St. George, Utah, the party turned southeast once again to return to Santa Fe and discovered, quite by luck, one of the few feasible places to cross the Colorado River in a stretch of several hundred miles. As a result of that expedition, a crude map was drawn of the region, with some errors, of course. But it was to be the only map available for nearly a century. And the center of it remained blank.

And so, in 1869, it was Powell's intent to explore that unknown portion of the canyon country, to map its terrain, describe its physical features and its geology, make observations on the flora and fauna. He knew that an overland expedition was folly, for it could take years to cross this terrible terrain. (Even today travel across this region is difficult and, in some places, impossible.) And so he chose the easiest of routes, the watery path, floating the Green and Colorado rivers. But he realized that this was not without its dangers.

As with many unknown places, a certain number of legends had sprung up about the Colorado River. Indians and trappers alike pictured the area as one inhabited by demons, a veritable death trap for those foolish enough to venture into it. There were stories of immense waterfalls and of places where the great river itself disappeared into the bowels of the earth. In his journal Powell wrote, "More than once have I been warned by the Indians not to enter this canyon. They consider it disobedience to their gods and contempt for their authority, and believed that it would surely bring upon me their wrath." Perhaps more than once during the trials of the journey would Powell be reminded of those words of warning.

Rio Virgin, Utah.
Photograph by
Jack Hillers,
ca. 1872.

Confluence of Hoback and Snake rivers, Wyoming

The party pushed off on the 24th of May and for the first several days found the Green to be a swift but placid river. There were no serious rapids in this stretch, and the men used these quiet days to good advantage, practicing handling the cumbersome boats. None of these men could be called an experienced river runner, and for some it was their first time at the oars. The boats were wooden, three of them 21 feet in length and made of sturdy oak, the fourth a 16-footer made of pine. They had been built to the Major's specifications back east and shipped by the new railroad to Green River. The party carried with them supplies to last ten months and such other equipment as axes, hammers, and saws to build shelters. They were prepared to encamp for the winter if necessary, and to continue their journey in the spring of the following year. Powell's men had been handpicked by him, and several of them were experienced wilderness scouts, hunters, or trappers. More than a few were also veterans of the Civil War. All were hardy, adventuresome men, capable of enduring hardship in the wilds, and this was precisely why Powell picked them. The trials ahead would confirm his wise judgment.

Powell's fellow adventurers, as he listed them in his own journal: "J. C. Sumner and William H. Dunn are my boatmen in the 'Emma Dean'; then follows 'Kitty Clyde's Sister,' manned by W. H. Powell [the Major's brother] and G. Y. Bradley; next, the 'No Name,' with O. G. Howland, Seneca Howland, and Frank Goodman; and last comes the 'Maid of the Canyon' with W. R. Hawkins and Andrew Hall." Good men all of them, in Powell's mind, but the adventure ahead would prove to be a strain on their relationship.

They encountered their first rapid in Flaming Gorge (named by Powell for its brilliantly hued walls), and from this point on rapids became more numerous. The more dangerous ones were either portaged or lined, that is, the boats were unloaded and then let down through the rapid by ropes or lines secured from shore. Such procedures were tedious

Powell expedition,
1871. First camp,
Green River, Wyoming.
Photograph by
E. O. Beaman.

and difficult, but frequently necessary. As their journey progressed and they became more experienced at negotiating the wild waters, the party chose to ride through rapids that might have given them pause earlier. Nevertheless, they did not seem to grow overconfident, always maintaining a healthy respect for the dangers. Particularly after the events of their 16th day on the river.

The Green has cut a succession of steep, narrow gorges in this region, beginning with Flaming Gorge, which merges imperceptibly with Horseshoe and Kingfisher canyons, then Red Canyon. Below Red Canyon the country is open and expansive, the river sleek and placid, as though gathering its energy for more violence. Then the walls close in again, steeper than ever, at what Powell named the Canyon of Lodore. Sixteen days out and soon after they entered this canyon, a particularly fierce rapid forced the party to pull in for a portage. But one of the boats was unable to make shore and was swept unprepared into the boiling rapid. It smashed on the rocks, and the three men, the Howland brothers and Goodman, barely escaped with their lives and only with the desperate help of the others on shore. The crew was badly shaken by this incident at "Disaster Falls." Later, and with great difficulty, they managed to retrieve some of their instruments (plus a keg of whiskey) from the shattered boat, but most of their extra clothes and a good amount of food were never recovered. The passage through Lodore proved long and arduous, for the river became a series of dangerous rapids. Time and again Powell and his men were forced to stop, scout the raging torrent, then decide to line or portage the boats. But despite the difficulties and the close brush with death, Powell was able to write in his diary that "this has been a chapter of disasters and toils, notwithstanding which the Canyon of Lodore was not devoid of scenic interest, even beyond the power of pen to tell. The roar of its waters was heard unceasingly from the hour we entered it until we landed here [the confluence with the Yampa River]. No quiet in all that time. But its walls and cliffs, its peaks and crags, its amphitheaters and alcoves, tell a story of beauty and grandeur that I hear yet —and shall hear."

Having reached the confluence with the Yampa, they camped for four days. It was a much needed rest, and this beautiful spot—named Echo Park by common consent of all in the party—provided the necessary respite for taut nerves. Although they had come but a fraction of their total journey, the men were showing signs of edginess. They had had many days of bad rapids, had lost a boat and some of their gear, and there was a general anxiety about what lay ahead. Having sampled some of the river's bad rapids, they knew that there could be much more of the same downstream. And perhaps some of it would be even worse.

For those who have never made a white water river trip, it may be difficult to comprehend the myriad dangers and the emotional and physical drain that those dangers can have upon the individual. There may be frequent stretches where the river booms along, broken by violent rapids, and boatmen must be ever on guard. A swirl of current or a barely submerged rock, coupled with just a moment of inattention, may spell disaster. In addition to the mental and physical fatigue created by these conditions, there is the psychological effect of spending many uncomfortable hours wet and chilled from the constant spray of waves and rapids. Finally, such things as food play a role in the morale of the party. The plan of the Powell men to supplement their rather tasteless dried foods with fresh game was not as successful as had been hoped. And so there was grumbling about the meals. Some felt that an inordinate amount of space had been allotted to Powell's survey instruments, space that might have best been used for carrying additional food for varying the diet. But Powell managed to keep the situation well in hand, and the men (with one exception) still retained much of their initial enthusiasm and excitement for the project.

On the 21st of June they pushed off again, finding the river greatly increased in volume by the added waters of the Yampa. For the next several days they encountered rapids again, stopping to line or portage the more severe ones. This section of the Green flows through what is now Dinosaur National Monument, slicing across the northwest corner of Colorado and into eastern Utah. When they emerged from Split Mountain Canyon, the

Snake River,
eastern Idaho

Flathead River, Montana

last of the steep-walled gorges in this section, they drifted out into a broad, open valley where the river meandered peacefully for 30 miles or more. They were once again in known territory, the Uinta Valley, where the Uinta River joins the Green. There were signs of Indian encampments here, and this country had been roamed by trappers and explorers. In fact, Father Escalante had crossed the Green here almost 100 years before. And about 30 miles up the Uinta River was a little settlement where the party was able to purchase more food and dispatch some letters to the outside world. It was here that Frank Goodman announced to Powell that he had had enough of the river's toils and dangers. He left to make his way back to civilization.

Soon after they pushed off from the Uinta River, stocked with additional supplies, they drifted toward unknown country again. Gradually, almost imperceptibly, the broad valley closed in, and walls began to rise near the river. For the next several days they floated through a canyon more austere than any of those preceding, a place where little vegetation grew along the 1,000-foot walls. By common agreement it was named the Canyon of Desolation.

In this stretch of the Green River, Powell identified and named four distinct canyons, though even he admitted that the demarcation between some of them was subtle and indistinct. In a hundred miles Desolation gave way to Gray Canyon, named for a blackish-gray seam of impure coal found there. Then came Labyrinth and finally Stillwater canyons. In the first three they found rapids as rough as any they had encountered, and at a particularly bad spot in Desolation Canyon they came close to disaster again. Starting down the tongue of a bad rapid, Powell discovered too late that at the bottom it made a sharp bend to the left, against a sheer cliff. Rebounding from the terrific force of the water off that wall, a transverse wave rolled the boat over, spilling Powell, Sumner, and Dunn into the foaming waters. The latter two clung to the overturned boat. Powell, buoyed by an inflated life vest (a necessity for a man with one arm on a wild river), was finally able to swim to the boat, and the three managed to beach it amid a pile of driftwood just in time to avoid being swept down into the next rapid. But two blankets, a barometer, some rifles, and the oars had been lost. After moving another half mile, they put in for the day to saw some new oars out of driftwood.

For days they were under severe strain and fatigued by many portages and linings of rapids. In some instances the river completely filled the canyon from wall to wall and made portaging impossible. Under these circumstances the party had no choice but to run the white water. Finally, however, they came to Stillwater Canyon and the river abated. They now drifted along with rising anticipation and perhaps some apprehension, for around any bend they expected to find the waters of the Colorado River (then called the Grand upstream from the confluence) merging with those of the Green. The men were concerned that this junction might be boiling with white water. On July 17 they reached the Grand and with relief found it a peaceful confluence. "Late in the afternoon," wrote Powell, "the water becomes swift and our boats make great speed. An hour of this rapid running brings us to the junction of the Grand and the Green, the foot of Stillwater Canyon, as we have named it. These streams unite in solemn depths, more than 1,200 feet below the general surface of the country."

Throughout the journey Powell had been recording numerous scientific observations on the general geology of the country, the various strata of the rock formations they passed, plant and animal life, evidence of prehistoric Indian ruins (they found many), and the relationship of these canyons to adjacent landmarks and mountain ranges. At every possible opportunity he was off exploring side canyons or climbing cliffs high above the river to make barometric readings and record elevations. Powell was thorough. The record he compiled on this and subsequent trips in this region still stands as an accurate portrayal of the geology and topology of the Colorado-Green River country.

At the junction of the Grand and the Green, the party camped for several days in order to make more of those observations. And of course it was another welcome rest for everyone from the toils of the river. Their disasters and near disasters had greatly depleted supplies. After sifting wet flour and separating the moldy from the unmoldy food, the men found they now had only two months' supply rather than the ten months' they had started with. Any further losses would put them in an extremely bad situation. Despite their plight, the days camped here were relaxing, and they managed to joke and find humor in much that was going on. They had no way of knowing that a few miles below them on the Colorado was the beginning of what they would come to name Cataract Canyon, the most fearsome part yet of their river journey.

A modern-day river runner has described the experience of running part of Cataract Canyon as something akin to being in a continuous five-hour car wreck. In places the walls rise sheer from the river, and the river itself drops steeply in elevation. Powell described one three-quarter-mile section as dropping 75 feet. Here, big blocks that have peeled off the steep cliffs above obstruct the river, and it churns and foams over and around these constraints with pounding fury. For eight days the river alternated between serenity and rage, mostly the latter. One day the party was able to make only three-quarters of a mile after several exhaustive hours of portaging the boats around a bad rapid. But finally the ordeal was over and, almost as a reward for their hard work, the party sighted bighorn sheep and managed to shoot two. They celebrated with a feast of wild mutton, the first fresh meat in weeks. Below Cataract they entered Glen Canyon.

. . . .

This river journey of Powell and his men might have been scored as a great symphony. Movements one and two have taken them through Flaming Gorge and the Canyon of Lodore, from the Uinta Valley into Desolation and Labyrinth and Stillwater canyons to the Colorado, reaching a musical climax in Cataract Canyon to end the second movement. To this point it has been an intricate score, alternating between *crescendo* and *diminuendo,* played with force and drama, *fortissimo* and *espressivo,* and with emphasis on percussion and brass, *allegro.* Now, the third movement, Glen Canyon alone, is played sweet and soft, *pianissimo,* a solo perhaps for flute or strings, *appassionata, legato.*

They found this 150 miles of canyon an enchanting place, different entirely in character from all they had seen before. Where most of the river chasms to this point had been

Denver Public Library, Western History Department

Powell expedition, 1871.
Photograph by
E. O. Beaman.

Overleaf:
White Cloud
Mountains and
East Fork of
the Salmon River,
Idaho

chiseled sharp and angular, Glen Canyon was soft, smooth, and gentle. And the river echoed this gentleness. Listen to some of Powell's descriptions:

The features of this canyon are greatly diversified. Still vertical walls at times. These are usually found to stand above great curves. The river, sweeping around these bends, undermines the cliffs in places. Sometimes the rocks are overhanging; in other curves, curious narrow glens are found. . . .

Other wonderful features are the many side canyons or gorges that we pass. Sometimes we stop to explore these for a short distance. In some places their walls are much nearer each other above than below, so that they look somewhat like caves or chambers in the rocks. Usually, in going up such a gorge, we find beautiful vegetation; but our way is often cut off by deep basins, or "potholes," as they are called.

On the walls, and back many miles into the country, numbers of monument-shaped buttes are observed. So we have a curious ensemble of wonderful features—carved walls, royal arches, glens, alcove gulches, mounds and monuments. From which of these features shall we select a name? We decide to call it Glen Canyon.

Past these towering monuments, past these mounded billows of orange sandstone, past these oak-set glens, past these fern-decked alcoves, past these mural curves, we glide hour after hour, stopping now and then, as our attention is arrested by some new wonder.

Lees Ferry marks the end of the third movement, Glen Canyon. Here, in a few years, a Mormon pioneer will operate a ferry across the river, but the demarcation to Powell is the change in the character of the geology once more. Now the canyon walls are sharp and angular again, a feature that signals a change in the river's character as well. The fourth

Powell expedition, 1872. Note Powell's chair strapped to boat. Photograph by Jack Hillers.

movement begins *allegro* as the river builds momentum in Marble Canyon. More rapids, some of them as bad, or perhaps worse than those in Cataract. At Soap Creek, for example, only a dozen miles or so into Marble Canyon, the sleek and placid river was torn into a frenzy of incredible roostertails and haystacks, some of them 10 or 15 feet high, a swirling madness of foam. Powell wisely portaged this and other bad rapids in the days that followed.

When they reached the mouth of the Little Colorado and the end of Marble Canyon, there was growing apprehension on the part of all the men. They had been on the river 81 days. Their clothes were shredded, food was growing dangerously short, instruments had been battered and lost, the boats had to be repaired frequently. Throughout their journey the pleasant interludes, such as Glen Canyon, seemed to be followed by ever more dangerous waters. Marble Canyon had been an ordeal; but worse, perhaps, was the growing awesomeness of the chasms, deeper and more forbidding than anything they had seen thus far, with walls that soared 2,000, even 3,000 feet above the river. The party seemed truly trapped in the bowels of the earth. And now, starting a little above the confluence with the Little Colorado, the country had become even more gloomy and foreboding, for they had entered that section of the Grand Canyon where the river has cut through thousands of feet of young sandstones to the very bedrock of dark Precambrian gneiss and schist. Now called the Inner Gorge of the Grand Canyon, these gloomy rocks, the very foundation of the earth and some billions of years old, make the river even more dangerous and swift, and the rapids more numerous.

Before starting on the last leg of their journey, the *finale* of the symphony, Powell wrote:

> We are now ready to start on our way down the Great Unknown. Our boats, tied to a common stake, chafe each other as they are tossed by the fretful river. They ride high and buoyant, for their loads are lighter than we could desire. We have but a month's rations remaining. The flour has been resifted through the mosquito-net sieve; the spoiled bacon has been dried and the worst of it boiled; the few pounds of dried apples have been spread in the sun and reshrunken to their normal bulk. The sugar has all melted and gone on its way down the river. But we have a large sack of coffee. The lighting of the boats has this advantage: they will ride the waves better and we shall have but little to carry when we make a portage.
>
> We are three quarters of a mile in the depths of the earth, and the great river shrinks into insignificance as it dashes its angry waves against the walls and cliffs that rise to the world above; the waves are but puny ripples, and we but pigmies, running up and down the sands or lost among the boulders.
>
> We have an unknown distance yet to run, an unknown river to explore. What falls there are, we know not; what rocks beset the channel, we know not; what walls rise over the river, we know not. Ah, well! we may conjecture many things. The men talk as cheerfully as ever; jests are bandied about freely this morning; but to me the cheer is somber and the jests are ghastly.

But the mood was not as cheerful as Powell suggests. The men had grown gloomy and dour. They had had their fill of adventure. They wanted an end to the journey. And perhaps their apprehension colored their ability to judge objectively, for they began to look for the end of the canyon much sooner than they had a right to. If their instruments had been in working condition, they might have found by elevation and latitude that they were still more than 200 miles from Grand Wash Cliffs, the end of the Grand Canyon and the beginning of the known world again, where Mormon settlements could be found. They pushed off.

The extremes of weather alone affected their morale. Here in the heart of the Grand Canyon the searing mid-August sun can be brutal, with daytime temperatures often exceeding 100 degrees. Then at night the temperature drops drastically, and with hardly any blankets left, the men were subjected to fitful, bone-chilling nights. To make matters worse, they spent a few miserable evenings in a cold rain, unable to find shelter. So austere was

Madison River, Three Forks, Montana

this place, so unfriendly, that at times the best campsites they could find were a few square feet among the dark, beetling boulders. There were no more friendly beaches and gentle alcoves. Worst of all was the expectancy; they had badly misjudged the remaining distance—70 to 80 miles in their estimation. In reality they had almost three times that distance to go. Time after time they thought they had reached the end of the gorge. But time after time the walls closed in again, deeper and darker than ever, and the men found themselves once more in gloomy confines, the world about them pinched off by the steep walls. The rapids were worse than anything they had experienced before, Cataract Canyon included, each new one progressively more dangerous. Their spirits rose and fell.

Even with today's neoprene rafts and skilled river guides, this section of the Colorado through the heart of the Grand Canyon is difficult and dangerous. Lava Falls, especially, is a heart-stopping, pounding madness of white water. Modern river runners are not infrequently flipped over and tossed into the water here by the immense waves.

As the days dragged on and the rapids worsened, the normally ultra-cautious Powell began to take chances, a recklessness brought on by desperation. Their food supply was shrinking rapidly and could be measured in terms of days now. Constantly they were forced to re-dry the remaining flour. And the bacon became so spoiled that they finally threw it away. To add to their misery, there was more rain.

Around the 20th of August their spirits were given a boost when they drifted out of the dark and gloomy Precambrian rocks and into colorful sandstones again. The canyon, though still immensely deep—5,000 feet or more here—opened up considerably, giving them a more expansive view of the country. Powell knew that the end of their journey, the end of the canyon, was marked by such sedimentaries, and thus they now felt that the end must be close. To further boost morale they made excellent time, one day making 35 miles despite frequent bad rapids. But it wasn't to last. On August 27 the river swung away from its general westward flow and ran south, deeper into the older formations and causing them much anxiety. Soon they were back into the dreaded rock, the gloomy granite. "It is with no little misgiving," wrote Powell, "that we see the river enter these black, hard walls." Then, almost to seal their fate, a few hours after they entered the granite gorge again, they came to a rapid that topped anything they had seen. They pulled into shore to scout it. From both sides of the river, they scrambled up steep cliffs to gauge the severity of the rapid and concluded that it was impossible to line or portage. To run it seemed suicide, but there were no options open to them. Except one: Leave the canyon on foot, hiking up the side canyon at whose mouth they now camped toward the north rim of the Grand Canyon, some 4,000 to 5,000 feet above. They reckoned that they could be no more than 25 air miles from the Mormon settlements along the Virgin River.

It was the lowest their spirits had sunk on the entire journey. Five days of food left. Should they risk their lives now in this impassable rapid, so close to the end of their adventure? Would they fare any better trying to get out over the unknown terrain of the side canyon and rims? Three of them, the Howland brothers and William Dunn, decided they would go no further by river. They would take their chances by heading up there into the wild, rugged country. Powell tried to convince them to stay, but they would not be dissuaded. They had had enough and would not risk their lives any further on this mad river.

On the morning of August 28 the trio departed, taking with them rifles to obtain food and leaving their share of the remaining stores for the river party. It was a sad parting, each group thinking the other foolish in its respective decision.

After the departure of the Howlands and Dunn, Powell and the remainder of the crew prepared for their watery ordeal. One boat was left behind and the remaining two were loaded with gear. To lighten the load and make the boats more maneuverable, some instruments and ammunition and some of Powell's precious rock samples were left behind. Then they pushed off, Powell's boat in the lead.

Much to their surprise, after the first series of waves the rapid proved to be no worse than others they had run successfully. Dripping wet, but elated with their success at negotiating this rapid, they drifted in the calm water below the waves, firing rifles to attract their three friends back to join them. But they did not appear, and as the river runners floated down the river they perhaps sensed that they might never see their friends again.

Later that day they came to another terrible rapid. One of the boats, straining at the rope tying it to shore, finally broke loose, tearing out a section of the bow, and went hurtling down toward the throat of the rapid. Bradley, inside the craft and caught off guard by this turn of events, coolly maneuvered the boat through the roaring waves and at the foot of the rapid stood up waving his soaking hat as a sign of victory. To Powell it seemed a gesture

Department of the Interior.
U. S.
Geological and Geographical Survey of the Territories.
SECOND DIVISION
J.W. Powell Geologist in charge.

Powell expedition, 1871.
Photograph by
E. O. Beaman.

Mule deer

Near Snake River, eastern Idaho

of confidence that nothing could stop them now. He too negotiated the bad waters successfully. And as confirmation of their turn of luck and superiority over the great river, that evening they left the dark granite and entered the friendly sandstones once more. The next day they emerged from the canyon at Grand Wash Cliffs, their ordeal over. They had successfully traversed the wild heart of the Great Unknown. No longer faced with the dangers and toils of rapids and white water, they drifted along on the calm river for another day. "The relief from danger and the joy of success are great," wrote Powell. "How beautiful the sky, how bright the sunshine, what 'floods of delirious music' pour from the throats of birds, how sweet the fragrance of earth and tree and blossom! The first hour of convalescent freedom seems rich recompense for all—pain and gloom and terror."

Shortly after noon on August 30, they came upon a Mormon settler fishing from the shore. Having been instructed several weeks earlier to watch for the bodies and wreckage of the foolhardy Powell expedition, he was amazed to find the adventurers alive, though bearded and bedraggled and browned by the sun.

Powell returned to a hero's welcome back east, but not before learning the shocking fate of the three who had left the party. They had succeeded in climbing out of the great gorge to the North Rim, there to be murdered by Shivwits Indians. It was a tragic note to an otherwise happy adventure for Powell and his men.

This journey was merely the beginning of a deepening involvement in the slickrock and sandstone country for John Wesley Powell. Two years later he returned to repeat the trip, making more studies of the geology, biology, and ethnology and bringing with him artists and photographers as well as scientists to record the features of the land. Powell urged the establishment of what was to become the United States Geological Survey and later served as its director for 14 years.

Though he remains relatively obscure in American history, no man worked harder to foster an understanding of the region he charted, urging a rational plan for the controlled development of the land based on a realistic appraisal of its resources. That Powell did not succeed in his own time was through no fault of his own. He was a victim of the national frenzy of expansionism and exploitation. In light of the new frenzy of exploiting the West, one wonders if he would be heeded today. It's questionable.

River of
No Return

I AM SURROUNDED by the Salmon River. Clad only in tennis shoes, swimsuit, and a puffy red life jacket, I drift along in the current. Not far away is my raft, my sanctuary should these waters get too rough. But for now I am content to lie back, allowing the buoyancy of the life jacket to suspend me in my watery medium, and to look up at a fish-eye view of the wilderness passing by: blue gash of sky between high, forested walls; beetling gray granite; ponderosa pine; glaring white beaches; flowers and greenery brushing the river; side streams scrubbing themselves white as they rush down to join my river. There are no sounds. No, I take that back. There are no *man-made* sounds, but there is the low hiss of the moving blue-green waters carrying me along. And as they move me, these waters also pull and suck at my body, twisting me in subtle eddies that belie the seemingly smooth, quiet surface. The phrase "living river" comes to mind and if, indeed, a river lives, then I am feeling its pulse. But it is a cold pulse, for even though a hot July sun blazes down on me, my body is numbed by the snowmelt of the Sawtooth, White Cloud, Boulder, Beaverhead, and Bitterroot mountains. Every side stream—and there are countless numbers of them—is teeth-aching, palate-numbing cold, and they add to the chill of these waters.

I have no idea how long I have been drifting in the river and I really don't care. Time? What is that? Obviously some utterly meaningless concept here in my watery womb. What I perceive is movement, sky, water, earth, color, form. But not time.

Suddenly I'm jarred from my trance by the voices of my companions calling to me. I lift my head to look and listen. From eyeball level the water is smooth as far as I can see. But somewhere downstream I hear a dull, ominous drone; an insistent sibilance that warns of a change in the river's mood. A rapid. A wild thought crosses my mind: What would it be like to run a rapid in life jacket alone, experience it at river level, and feel the full violence of the river as it foams and churns its way over the rocky constraints? The idea is intriguing, but as we drift closer I begin to have second thoughts. Maybe I ought to start on a smaller rapid or some riffles before taking on a large one. Convinced by my own logic, I scramble, somewhat relieved, onto my raft and once again take over the oars.

The rapid turns out to be a minor one, typical of the hundreds of places where the river is pinched together by canyon walls, forcing it to increase velocity and to tear itself apart as it makes a noticeable drop in elevation. We pitch and toss over some nice waves, and soon the river spreads itself into a placid sheet. And again we drift quietly along.

This is the second (or is it the third?) day of the trip. My companions have come from widely scattered parts of the country: Los Angeles, New York, Chicago, Dallas. All meaningless abstractions here. Our raft is part of a flotilla of three rafts in a party run by

Snake River, Swan Valley, Idaho

my old friend Jim Campbell, who owns and operates Wild Rivers Idaho. Jim is at the oars of one raft, Cort Conley mans another. Mine is the third and smallest of the three, and I prefer it that way.

This is a totally new experience for me. I've been on the Salmon many times before and on other rivers as well, but this is my first time as a boatman or guide manning the oars, responsible for the safety of my passengers. And added to the weight of responsibility is the thought that my own two children, Jean Anne and Scott, accompany me on this trip.

But such thoughts are not really burdensome in this country. In the spell and grip of this river nothing really matters. It feels good to be home again, back on the Salmon, *my* Salmon River. What is there about this place? Why am I drawn back here again and again? I dislike being analytical about it, yet there is something about this Salmon River country. It's a gentle wilderness, soft and welcoming, with no sharp edges or harshness. I have a feeling that I could step off my boat at some convenient landing, bid the others farewell, and settle down here easily. An abundance of trees would provide me with wood for shelter. And game is plentiful. Thus far we've seen many deer and there are countless elk in the high country. A struggle for survival? Seems more like living in the lap of luxury. But maybe I deceive myself, maybe this land is not as gentle as it first appears. People have tried for a hundred years or more and they've been beaten back, held in abeyance. All along our 90-mile journey is evidence of attempts to tame the country: Crofoot Ranch, Allison Ranch, Campbell's Ferry Ranch (no relation to Jim, our head guide), the Moore place, James Ranch, Shepp Ranch, the Polly Bemis place, and that's just about it. There was ample incentive here: gold. And remnants of the diggings are still around: Nabob Mine, Painter Bar Mine, Golden Anchor Mine, War Eagle Mine. Farther up in the high country were Leesburg, Grantsburg, Warren, Dixie, the boom towns that are now turning to rust and dust, decaying along with the remnants of the mines. For the gold seekers, this country was a temporary home, never intended to be more. But others decided to stay and they built with love and care their homes of immense, handhewn ponderosa logs, painstakingly cut, squared, and fitted one by one. Charlie Shepp's place took two years to build. With care it'll last another two centuries or more. And even the Moore place, abandoned for several decades, still has a sturdiness and solidity that seem almost geological.

But I digress. To get back to my original thought, why hasn't this land been tamed like the rest of our country, or at least like the rest of the West? Why aren't there roads, highways, cars, suburbs, factories, condominiums, trailers, hotdog stands, tramways, amusement parks, people, pavement, and pollution? Was this land so overpowering, the early pioneers so weak, that it resisted being subdued? Is this place both Valhalla and Hell? The answers to these questions seem to lie in examination of the total scope of this place, for, being confined to the depths of this great canyon—the second deepest on the continent, I'm told— we are cut off from the size and character of all this wilderness. Aside from occasional glimpses of the high country some 4,000 to 5,000 feet above the river, we have no feeling for the vastness, the great extent of the country surrounding us.

I remember once flying over here on a commercial flight to Seattle. It was evening, and the last flickering colors of an incredible sunset still lingered on the horizon when our pilot announced that we were over the "great Salmon River wilderness." I detected a curious mixture of pride and awe in his voice. All of us strained, of course, to see into the blackness below, but nothing was visible. Then it struck me: The blackness, that great expanse devoid of any sign of civilization—not even any lights—was the most obvious mark of a vast wilderness. We flew on for a long time at 30,000 feet and 500 miles per hour, and still no lights, no sign of roads or cars or towns, and I wondered how many places are left in this nation so free of man's presence.

As I learned later, this region through which our rafts now float is the largest expanse of wild country this side of Alaska, roughly 6,000 square miles and still roadless and wild. So far. And three of the continent's last wild rivers flow here: the Selway, the Middle Fork of the Salmon, and these waters of the main Salmon that bear us swiftly along. So perhaps it wasn't the harshness as much as the vastness that insulated it all from man, and still does to some extent. This incredible up-and-down, creased and wrinkled country of deep gorges, turbulent waters, and thick forests still has the ability to overwhelm man. It resisted Lewis and Clark and a good many others who came afterward. But one wonders how long it can continue to hold out.

Drifting along and discussing these thoughts with my passengers nearly causes me to miss Jim and Cort, who are pulled into a sandy beach around a bend on the right bank. I pull hard on the oars to slip out of the main current and into a back eddy where I can nose the boat onto the sandy shore.

Lunch break, we discover as we step ashore. The heat is fierce, and we seek shelter under the cool, graceful limbs of a ponderosa pine. I try one of Jim's Salmon River Specials, a sandwich comprised of salami, cheese, tuna fish, mustard pickles baloney mayonnaise, and topped with a generous layer of peanut butter, guaranteed by my chief guide to give exercise to even the most indolent digestive tract. This gastronomical nightmare is washed down with delicious icy cold water dipped from a nearby side stream.

To work off my lunch I decide to have a look around. The little stream is edged by moss-covered rocks, and ferns grow in profusion. An Eden that needs to be explored. As I push my way upstream I discover that it's no easy task, however. It is a jungle of thickets

Denver Public Library, Western History Department

Light House Rock,
Green River.
Photograph by
E. O. Beaman,
1871.

Camas National Wildlife Refuge, Lemhi Mountains, Idaho

Trumpeter swans

and undergrowth, dense and lush. More reminiscent of a rain forest than the Rocky Mountains. It's a curious fact that there are many such micro-ecosystems along here. They exist along stream courses in protected side canyons and gulleys where there is a moister environment. These places are totally unlike the typical forests of the Rockies, which are characterized by a general sparseness of vegetation: slim, spare lodgepole pines, open ponderosa forests interspersed with parklike meadows, or alpine fir growing in clumps near timberline. A few years ago I discovered one of these Edens in an obscure, rarely visited side canyon here along the Salmon, a place so incredible, so beautiful, that each time I visit, it seems unreal. It lies some distance from the river, and getting there is a terrible bushwhack. I doubt that more than a few people have ever seen it. The walls of the canyon close in, and a little waterfall plunges over a lip of rock into a deep, secret pool. The overhanging walls of this grotto are covered with mosses and ferns, and the dense vegetation closes it all in, shutting out sky and sun. The carpet of mosses and ferns and flowers on the floor of this place seems so delicate that I walk very carefully, on tiptoe, to avoid disturbing anything. It is so lush, so tropical, that I half expect (and half hope) to find some lovely Tahitian girl sitting by the quiet waters of the pool.

With visions of discovering another such place, I stumble along this little stream. I recognize several plant species that are totally out of place here in the Rockies: maples, with their palmate, serrate leaves; and trillium, the delicate three-leafed, three-petaled flower. I would stake a case of my favorite beer that no trillium grows in all of Wyoming or Colorado.

As I'm about to step off a moss-covered, decaying log, I'm startled out of my thoughts by a high-pitched buzz. Having encountered many rattlesnakes in my travels, I recognize the sound instantly, and my response is learned: Freeze until you locate him. He's not difficult to find, lying coiled next to a broken extension of the log I'm on and about six feet away. Undoubtedly he heard me coming from a long way off, with my thrashing and crashing through this dense brush. But he chose to wait until an encounter was imminent before sounding off. As I stand watching him, he begins edging away, head and neck pointed toward me, but body slowly moving toward the protection of denser brush. Searching around, I find a broken branch nearby and I use it to halt his getaway, poking him gently back into a firm defensive pose. His buzzing rattle is loud and insistent. I hunker down on my log to study him. About six feet of space and a hundred million years in time separate us. He slowly flicks out a black, forked tongue and leaves it suspended in space a few moments. Tasting the air. Slowly his vibrating tail comes to a stop as I sit quietly, making no noise or movement. He eyes me cautiously through dusty, unblinking eyes, the elliptical pupils of which are dilated somewhat by the subdued lighting here. About two feet in length, his muscular body is sandy brown in color with dark brown blotches, or saddles, that are spaced regularly along his length and extend down his sides. Pacific coast rattler, I decide, or possibly a prairie rattler. No expert, I find it difficult to distinguish the two. No matter. He, possibly, is trying to classify me as well. This particular area is remote enough that I may be the first human he's ever encountered. "My God," he's probably thinking, "the gang back at the rocks is never going to believe this."

As a test I raise my arm quickly, and just as quickly he sounds his rattle again and tenses his body. When I make no further movements, he silences his tail in a few moments. It occurs to me that this would make a beautiful photograph, but here am I, dedicated photographer, without a Nikon or Rolleiflex shooting iron anywhere in reach. Finally I stand up, ready to head back to my companions, and he once more sounds his alarm. "That's all right, old friend," I tell him silently, "I mean you no harm. Thanks for letting me visit." I back away, then turn and retrace my thrashing path, cautious now that maybe a brother or a cousin of my friend back there might be lurking where I might step on him inadvertently.

Back at the raft, everyone is anxious to push off, move on, see what lies around the bend. I mention my encounter with the rattlesnake, and several people are horrified that I didn't kill him. "But why?" I ask. "He did me no harm. And besides, I was the intruder into his domain."

It gives them something to think about as we push off. Or perhaps they wonder what kind of a nut they have for a guide.

The river once more pulls us along. In a few more bends, we approach one of the major obstacles of the river, Salmon Falls, which is not really a falls at all, but an abrupt rapid. Those first few moments before slipping down into the watery fury are always a bit tense. Ahead, the smooth surface of the water ends abruptly at a thin, white line. Beyond, the dancing tips of white waves can be seen, and the sound is that of several freight trains rumbling through the gorge. I hang back, letting Cort then Jim run through. Finally I maneuver to start down the slick tongue into the rapid. Everything happens quickly, and there is little time to think, let alone react. After the first chute we plunge into some huge stationary waves that curl back on themselves and toss the raft around like a cork. As we pass swiftly through the middle of this madness, one curler reaches into the back of the raft and floats my friend Jimmy Collier clean out of the boat, then washes him back in, all in the same instant. In a moment we bounce madly through the tail waves and out into the placid water below, where Jim and Cort and the others all wait. Everyone is soaked and excited, still high from that hot, coursing jolt of adrenaline, and the raft spins slowly in the quiet water as everyone recounts the experience. "Did you see that wave that . . ." "I almost went over when . . ." "Hey, Jimmy, you're supposed to stay in the boat." Then we quietly settle into the peaceful mood of the river again. A hot sun blazes in the late afternoon sky, and the river becomes a slick glare, painful to look at. In a few miles we pull into Bruin Creek to make camp for the night, and the shade of the forest is welcome respite from the inferno. It is not long before the sun slips behind the high, verdant hills hemming us in, and the canyon is then steeped in cool shadow. There is no real sunset. Darkness simply sneaks in and steals away the remaining light. We sit around the campfire sipping coffee and digesting an immense meal, reading the mystic messages of our fiery oracle. Someone tells a story. Another relates a bit of adventure from today. Jimmy Collier, an up-and-coming folk singer from New York, plays some quiet songs on his guitar. The world outside and beyond these canyon walls could well be engaged in nuclear holocaust or a stock market crash or any number of calamities political or economic. But here the most troubling thought is choosing a comfortable spot on the sandy beach for your sleeping bag. As I lie there before dropping off, I hear Jean Anne and Scott not far away talking excitedly from their sleeping bags about the day's happenings. I resist a fatherly impulse to suggest that they should get to sleep. To hell with it. Let 'em stay up and count stars all night long. The world needs a lot more star counters.

· · · ·

"How did it come to be called the 'River of No Return'?" someone asks next morning at breakfast. Cort Conley relates some of the river's history.

The first recorded encounter of white men with the Salmon was that of Lewis and Clark in 1805. After following the Lemhi to its confluence with the Salmon, William Clark named this "handsom river" after his partner. The explorers had hoped it would provide them easy passage to the Columbia and thence the Pacific shore, but after following the river for a way into this gorge, Clark found the waters "foaming & roreing thro rocks in every direction, So as to render the passage of any thing impossible." They gave it up and headed north to find an easier crossing of the northern Rockies.

Fur trappers? Yes, they wandered through here, though not as extensively as in other parts of the Rockies. Little mention is made of the Salmon in trappers' journals. It was a long way to the Salmon from the Rendezvous areas of the upper Green River, most of the journey crossing the heart of hostile Blackfoot country. In 1832 a party of trappers attempted a descent of the river by canoe. Two of them drowned in the ordeal, which took some 30 days to complete. Records are a bit sketchy of other attempts. In the 1870s one Johnny McKay was apparently successful at it, and on the previous afternoon we had stopped at a lovely hot springs where McKay had left an inscription on a rock. In 1896 Harold Guleke, called "Captain" by his friends, successfully negotiated the wild river from Salmon city to Riggins. Captain Guleke began regular commercial trips by large

113

Gallatin River, near headwaters of the Missouri River, Montana

Colorado River, Grand Canyon, Arizona

wooden barges, some up to 32 feet in length, built in the town of Salmon. He carried supplies to the mining town of Riggins, occasionally continuing on to Lewiston, some 150 miles on down the Salmon and Snake. He once ran all the way to Portland, Oregon. It was ostensibly a commercial venture, but some feel he ran these trips during a 30-year period purely for the adventure and love of the river, for he could hardly have made much money. At the end of each trip, he sold his boats and barges for their lumber; currents made the return by water impossible, and he obviously could not transport the boats over mountain and valley several hundred miles back to Salmon city. And thus it seems in this period the Salmon became the "River of No Return."

Although his boats did not return, Guleke himself did many times. He floated the wild waters as late as 1934, when he was nearly killed in Salmon Falls. But the Salmon remained relatively obscure to the outside world until the National Geographic expedition of 1936. Even with that type of exposure, however, it wasn't until the outdoor recreation boom began in the 1950s that significant numbers began running the Salmon.

• • • •

Breakfast and history finished, we pack our gear and load it into the boats. The blue coolness of early morning is dissipating rapidly as the sun blasts over the hillside opposite us. Days in the canyon are shortened considerably by the pinch effect of the high hills and canyon walls that close us in, a blessing in some ways, for in midsummer it gives some relief from the blistering sun. As hot as it gets here, though, it still doesn't quite compare with the occasional 120-degree temperatures of Hells Canyon. That gorge was aptly named. But when you're on the river the heat can be dealt with effectively; when it gets to be too much you simply peel over the side and into the cold waters. It's something like a sauna: You bake until you can't stand it, then quench your body in the icy water. Does wonders for the circulation.

Back on the river again, we drift along on quiet waters. Side streams plunge out of green grottoes to join us. Backlit ponderosa pine becomes a blaze of long, silvery needles. Yellow arnica, blue lupine, and red Indian paintbrush wave at us from the shore. There is no doubt about it, I think to myself, the Salmon is my favorite river. For pure adrenaline, the Colorado can't be beat, of course. It is a raging monster, beautiful in its own way, with that eyeball-blasting color of the great slickrock country. Being a part-time desert rat, I dig that country, as they say. Then there's the Snake in Hells Canyon, sister of the Salmon, but different; more austere, dark, sinuous, moody, like something described by Tolkien. I love that country too. Even the Middle Fork of the Salmon is different in character from the Salmon proper, swifter, rockier, bawdy and brawling. Then there are the Selway, the Green, the Yampa, the Owyhee, and others. All beautiful, each distinctive. But I keep coming back here. If I believed in reincarnation, I would have to assign a former life to this place, so strong is the feeling of attachment.

The current bears us along at a good pace while eddies spin the raft slowly, giving a changing panorama of sky, trees, water. Somewhere along a rocky wall, a canyon wren whistles its descending, melodic song. We whistle back. It answers. Then some rapids, small ones, but enough to get us wet as we bounce through the waves. Better ones are ahead today: Bailey Creek, Split Rock, and Big Mallard. Or B-b-b-b-big Mallard, says Cort Conley in mock horror. I remember that one well from my first trip on the Salmon. Just Jim Campbell and me on a scouting trip prior to Jim's starting his river business on the Salmon. We were rowing one of the 28-foot rafts with two sets of oars, Jim on the back, me on the front set, with him calling the shots: "Pull left," or "Pull right." Starting into the tongue of Big Mallard, we hadn't seen that big rock sitting down there, water washing over it with tremendous force and creating a huge hole in its downstream side, a hole big enough to swallow us, raft and all. I was enjoying the waves as we plunged through the rapid, when Jim suddenly yelled, "Pull back. *Hard!*" Detecting the note of panic in his voice, I leaned back on the oars with everything I had. Trouble was, I did so at the moment the front of the raft was riding up on the crest of a wave. My oars bit into empty air and I fell over backward, simultaneously catching the full force of a wave breaking over the front. We missed that rock and its hole, but not by much, and I still don't know how. Almost bought the farm, as they say on the river.

"Hey, Norton," a voice from the present shakes me out of the past. It's Jimmy Collier.

"Yeah?"

"What's to keep someone from building a dam here, or logging these forests, or building roads?" Jimmy has been an environmental activist with Pete Seeger on the Hudson River sloop the *Clearwater*.

"You and me and the rest of us here," I say, waving my arm to take in the others in the raft. I go on to explain the status of this river and the country around us. The river is being studied for possible inclusion under the Wild and Scenic Rivers Act of 1968, which, if achieved, would assure reasonably long-term protection against such things as dam building. But the surrounding country, well, its fate is a little less certain. On the left bank as we float downstream is the Idaho Primitive Area, 1.2 million acres in size, on the right the Salmon River Breaks Primitive Area, some quarter of a million acres large, the classification of both somewhat vague and tentative. The U.S. Forest Service is in the process of recommending to Congress that they be reclassified as wilderness areas for permanent protection. But the Forest Service is recommending far less land area than needs protection, and big developers want it all declassified, thrown open to development. Like logging.

I let that sink in as we drift along beneath these thickly forested slopes that have never seen ax or chain saw. This country has huge herds of elk and deer, sizable numbers of bears and bighorn sheep, the rare mountain goat, the rarer mountain lion. In fact, I point out, this is the last real stronghold of the cougar. All the wildlife will be adversely affected if this country is opened up for logging. And the river especially will suffer.

"How?" someone asks.

It won't be the same river as we see it. Roads and other developments go hand in hand with logging. But most important, no river runs pure and clean in logging country.

They think about it.

"But don't we need the resources?" someone else asks. The inevitable question, one that I've answered, or tried to answer, many times before. Yes, I explain, trying not to preach. I suppose we do need the resources *if* we continue this madness we've created—expanding population, reckless consumption, one-time, throwaway, wasteful use of resources. But, I go on, we've preserved only about one percent of our country's land area as wilderness areas or national parks. And if we manage to ultimately preserve this and other areas in need, we *might* end up with 4 or 5 percent. Now if we can't make it on 95 percent, then how in the world is this remaining 5 percent going to save us in the long run?

They hadn't thought about that.

"But what about the cost?"

Good question. It is going to cost. Everything worthwhile seems to have a price nowadays. But we can get this for a bargain. We can have it for a slightly lower thermostat setting, a few more miles per gallon, the use of more recycled products, a more miserly attitude in the use of resources, the development of . . .

"Okay, Norton, we're sold. What do we do?" asks Jimmy the Activist.

Yell. Scream your bloody heads off. Write letters to your Congressmen asking enactment of a bill for a 2.3-million-acre River of No Return Wilderness Area. Because if you don't act, you know that the big corporations will, lobbying like mad to get their hands on this place and sell us down the river, as they say.

End. Sermon delivered, I say as we race through some choppy waves, then pull into a backwater eddy at the white beach of Bargamin Creek for lunch.

Of the hundreds of side streams that meet the Salmon, Bargamin Creek is one of the largest. At this stage of runoff it runs strong and clear, and near the junction with the Salmon are piles of large, white boulders polished by both the Salmon and Bargamin Creek during flood stage.

You know, I say to my raftmates, I don't mean to belabor a point, but did you realize that the Salmon River is one of the few sizable rivers left in the nation undammed

Bryce Canyon
National Park,
Utah

Virgin River, Zion National Park, Utah

over its entire length? And whether we, *homo sapiens,* like it or not, an unfettered river takes care of itself very nicely by flushing itself out seasonally. This lovely beach is a product of that. Each June at high water the Salmon floods through here, a raging torrent, and it sweeps the channel clear, then deposits fine white sand as it slowly drops in velocity after the peak of runoff.

Across the way, high-water marks can be seen some 30 to 40 feet above the present level. Driftwood logs and whole trees rest on rounded boulders near the line of forest along the upper edge of that mark.

"Must really be something to see at flood stage," someone remarks.

My thoughts wander back to a trip on the Snake River several years ago in late May and in a year of early runoff. At the confluence with the Salmon, we found it a rampaging, coffee-colored torrent, carrying huge trees ripped from shorelines upstream, perhaps even from these very shores where we now sit. It was an awesome spectacle and an impressive display of a river's power.

After lunch we push off again and about a mile below Bargamin run Bailey Creek rapid. Left of center, I say to myself, mentally gauging the run as we drift up on it. Follow the slot, pull away from the rock on the far left. We plunge into the rapid, smack a wave, buck another, pitch and toss into the tail waves. A good run, I congratulate myself. Now the river picks up in tempo. Around the bend, more white water. Split Rock rapid and several smaller ones are run in quick succession. Then, sooner than I expect, we come to B-b-b-b-big Mallard, my nemesis, my jinx. The river bends sharply to the left, making it impossible to see the lower section of the rapid. We pull into shore on the left bank to scout it.

On the grassy slope above the water, Jim and Cort and I look down on the madness of rock and foam. That big rock is fully out of the water now, left of center, but all the water from the right side sweeps toward it and piles up, making a run down the right side out of the question. On the left a chute of water pours down between the big rock and another just off the left shore. That's our run, says Jim, the narrow slot between the two rocks. To me it looks pretty thin. Might scrape the paint job on this one.

Back in the boat, sweaty palmed, I have that cottony taste in my mouth, though I try not to let on to my raftmates that I'm nervous. If I were alone I would be less worried, but being responsible for the five others in my raft gives me some concern about the run. Hitting either of those rocks could flip us or swamp the boat in an instant.

We watch first Jim then Cort push off and start down that left-hand tongue. I don't envy their maneuvering those big rafts. If they jam up sideways in that chute, they're in real trouble. They round the bend and slip out of sight.

Then we're off. The water is slick and smooth here, but soon the current catches hold, pulling us irresistibly toward that roaring maelstrom. Facing midstream I pull away from a rock, then spin us around quickly to face the left shore. I stand up to get a better look down that ramp. Below the rapid wait Jim and Cort, safely through now. Deep breath. All right, Norton, this is it, I think to myself, my mouth so dry I couldn't spit if I had to. We pick up speed and begin the plunge. Pull with the right oar to straighten us out, point us down the chute. Too far. Pull left oar. It hits a rock, popping out the oarlock!

"Jimmy!"

Collier, seeing what has happened, leaps forward to help me put the oarlock back in, while I try to maneuver with the other oar. Water roars all around us as we plunge headlong and out of control toward the slot between the rocks. It seems like an eternity before the oar is back in, but then it's too late as we race between the rocks. I pull the oars back to keep them from hitting, a wave breaks over the front, drenching everyone from bow to stern. We shoot over another wave, into a hole, then bounce over the tail waves and into quiet water. Made it!

"Hey, Norton, what are you, some kind of a showoff? How 'bout using both oars next time?"

"Good old One-Oar Norton. What do you do for your next act?"

That one was a little too close, I think to myself, still giddy from the adrenaline high. That's twice now in B-b-b-b-big Mallard. Damn rapid may get me yet. But soon the roar diminishes, and we drift along in quiet waters again. The afternoon flows on as quietly and as gently as the river. Camp that night is in the grassy meadows of Rhett Creek, site of an old homestead long abandoned. We feast on wild asparagus, watching the blue of evening slip into the canyon. Jimmy sings on into the darkness, but I turn in early, tired from the day's run.

·   ·   ·   ·

From observation of both myself and others, I've found that people go through three mental stages on a river trip. First of all, there is apprehension, fear of the unknown, misgiving. (Do I really want to go through with this? What the hell am I doing here?) But this is overcome by the concurrent sense of expectancy, excitement, adventure. This first stage, the shortest of the three, lasts anywhere from an hour to a day.

The second stage is the longest and is characterized by happy oblivion, or what could be called the I-just-don't-give-a-damn-about-the-state-of-the-outside-world syndrome. You are a captive of the river, and it sets the pace of your journey. Nothing you can do will speed it up, so you might as well get used to it. Moreover, you become attuned to the river and the land. If I may risk becoming a bit esoteric, there's a feeling of oneness with the country, as though for the first time you really understand man's ties to the earth. Water, sky, earth, air—all take on new meaning. This stage is also dominated by what I call River Time. Like those sweet childhood days of summer, the minutes, hours, days blend into a crazy whirl of impressions with no sense of order, no *need* for order. Let's see, when did we hike up to Rattlesnake Ridge? Two days ago? No, three. I think. And what day is today? Thursday? Who knows? Who cares?

But stage three is insidious. It begins to creep in subtly by the latter part of the trip and culminates in the sudden realization that it will soon be over. Characteristically there is a feeling of panic, sadness, and the need for cramming even more into the brief hours left. You look around at the people who have shared campfires, stories, storms, rapids, beauty, and danger with you and you read some of the same thoughts. We must write. Yes. And if you ever get to New York Chicago Dallas L.A. Springfield Washington Denver you must stay with us. Yes. We must do this again next year. Yes. And the year after. Yes. But the gut feeling is that this was a one-time, one-shot thing. No repeating. River friendships and river love affairs are intense but brief.

·   ·   ·   ·

For two days we drift progressively deeper into stage three, the river smooth, roiling, quiet, noisy, the trees and flowers and beaches and rocks slipping by too quickly, each needing to be studied, embraced, cherished. Having been through it all many times before, I should be prepared for what happens at the end, what my river friends call the Flaming Re-Entry to Civilization. But I can never quite ready myself. It is culture shock in its rawest, most brutal form and it numbs the mind, dulls the senses. Trying to soften the impact for Jean Anne and Scott does no good, and there is a teary sadness as they watch the great rafts slowly deflate on shore like some tired, friendly creatures that have carried them through a great fairy-tale adventure and then expired. We will come back again, you know. Yes. I promise. Yes. It is a promise that I will keep. I must.

River People

WHEN BUTCH CASSIDY ROBBED the bank in Telluride, Colorado, in 1889, he made his getaway south, over Lizard Head Pass and down the Dolores River. Then, according to stories in these parts, he headed over the mountains to the West Dolores River, where he stopped overnight at the little town of Dunton for lodging and meals and perhaps a dip in the pleasant hot springs. There was no way for the citizens of Dunton to know that they were sheltering an outlaw and his gang, for there were no telephones or modern means of communicating the calamity at Telluride.

Today Butch and his gang might have a tougher time of it, what with closed circuit television and electronic surveillance, shortwave radio and speedy transportation. But if he could make it as far as Dunton once more, he might again relax in the hot pool for a while, free of troubling thoughts about the law. You see, there is still no telephone in Dunton.

Emilio Roscio, called Millo by his friends, was born here in Dunton 16 years after Butch passed through. In 1905 the town was a busy little community of . . . well, no one seems quite certain how many. Five hundred, maybe a thousand. Hardly a boom town of the proportions of Central City or Leadville or Cripple Creek, but substantial enough for the citizens here. Perhaps it wasn't gold alone that kept people here, for this upper part of the West Dolores River, with Dunton alongside, sits in the heart of some lovely country. Steep hills rise from the valley. Not big mountains like those that tower over Telluride and Ouray, but big enough. And from a little ways downstream can be seen that freakish monolith to the north called Lizard Head, an obscene rocky finger poking into the clear blue sky and seeming to guard this valley against the outside world in a symbolic sort of way. The mountainsides around the valley are pretty heavily forested. A lot of spruce. Lower down it's mostly aspen, and it blazes a brilliant yellow in the fall. In some ways this place is more reminiscent of the White Mountains of New Hampshire than of the Rocky Mountains.

Dunton is a pretty lonely place now compared to when the mines operated, and judging from the very few signs of mining residue, they apparently were never big operations. The West Dolores River flows clean and clear, meandering through open willow flats, then into spruce forests. They say the fishing is great.

Millo has seen a lot of changes in his time, but here along the West Dolores things have changed a lot less and a lot slower than in the "outside" world.

"It was about 1920 when the first automobile come up the valley," he says, adding, "Pretty rough trip in those days." I can well imagine, for even today the gravel road leading in here is hardly a model of smoothness. But more intriguing, I think to myself, is the fact that the auto made its debut here almost 20 years after it showed up in other parts of the Rockies.

Even with the coming of the auto, however, the way of life didn't change much along the West Dolores River; people still depended largely on the tried and tested methods of transportation. "It wasn't so terribly long ago that the mail came over the mountain by wagon from Rico. Three times a week, even in winter," says Millo. "Used to cut willow stakes to mark the road up there when the snow blew over it," he says with a sweep of his hand indicating the high-country pass to the north of us. Having spent many years as a hunting guide, he speaks of places in these mountains with a sure, intimate knowledge.

The mines here were operating full bore at the turn of the century, but by 1910 they had closed, played out. Dunton—named for an early prospector from Maine, Horatio Dunton—diminished in numbers as people moved on. But some stayed, homesteaded the 160-acre ranches along the river, and lived comfortably, raising some crops and beef. For the next few decades, the mines reopened intermittently as money was raised for new ventures. But sooner or later both the money and the ore ran out. The mines closed for the last time in 1941.

What was Dunton like way back, and especially when the mines were running? "Pretty rough place at times," says Millo with a quiet sort of pride in his voice. "We were quite a ways from the law, you know. You kinda had to be rough right along with 'em to survive." He says it with a chuckle, and the rough lines of his face become creased in a smile that speaks of a bit of mischief of its own. Saturday nights in downtown Dunton were for celebrating, washing away the toils of the tunnels or ranches. And if some exuberant cowpoke or miner rode his horse through the little saloon occasionally, why it was all in fun and no one got hurt. Well, hardly anyone.

From outward appearances, Dunton hasn't changed a great deal, at least not so's you'd notice. Today the little collection of log buildings is privately owned and operated as a resort. Not the big free-wheeling kind of resort you'd find in other places, with fancy restaurants, curio shops, golf courses, and the like. There's none of that nonsense here, not even any signs advertising the fact that you've arrived, and hardly any improvements have been made to the lovely, weathered log cabins. It's pretty informal. You just sort of show up, sign in (if they have a cabin available), eat your meals ranch style at a long table in the hotel dining room, soak in the hot pool, fish in the river, or sip a drink in the old bar where Butch and his gang sat. (As a concession to modern tastes, they do have a limited selection of liquor other than the old standby, straight whiskey: scotch, sometimes gin. But you'll get a strange look if you ask for a Harvey Wallbanger.) The hotel and cabins have never seen a coat of paint and are weathered into that lovely brown-gray patina of hundred-year-old wood. And if Paul Cummins, the young sandy-haired, sandy-bearded manager of the Dunton "resort" has his way, things'll stay just as they are.

In contrast to Millo Roscio's longevity in the valley, Paul is a newcomer, having been here less than a year, and coming by way of the circuitous route of Boulder and Denver and Breckenridge and a few other places, all inferior to this spot. He has found it easy to slip into the peaceful tempo set by the gently meandering river, and one has the distinct impression that Paul has lived most of his life here.

"I feel downright defensive about this place," he says. "I don't want it to change. The people who come here put up with the lack of conveniences because they're sensitive to the country. They're mellow people," he adds. "If we can make a go of it by just word-of-mouth communication, why, so much the better. I like to tell people 'Don't come here unless you've got your head together about such things.' We're sure not a Holiday Inn."

To Paul the ideal situation would be for Dunton to become completely self-sufficient by tapping (with care) the geothermal energy sources that feed the hot springs, growing crops, and achieving a level of social and economic activity that would allow people to realize their full creative potential. He envisions attracting minimal numbers here with such low-impact activities as ski touring in the winter and hiking and backpacking in the summer. "I think business here can be handled more artfully than in other places. At least, we have a chance to try it."

Near Hells Canyon, Idaho and Oregon

Grand Canyon of
the Yellowstone,
Yellowstone
National Park,
Wyoming

But what about the twin spectre of Change and Progress that looms darkly on the horizon? I wonder aloud about the growing influx of people, new roads and highways, more industry. Over in Telluride, for example, people worry about the new ski area development turning their town into another Aspen or Vail.

Paul: "I'll resist all efforts at growth and progress here."

For Millo, who has been through both the good times and the bad, it's not as clear-cut. "I'd hate to see things get messed up around here," he says. "I like it the way it is. I don't think you'll find prettier country or a nicer river." But there's also a detectable, gut-deep feeling of needing a little more security, perhaps a few more jobs, a more stable economic base, just a few more people. Already more and more people are arriving, fleeing the mess they created in other parts of the country. He laughs at the paradox. "Yes," he says, "and the more of 'em come, the more of a mess they make here. I just don't know what the answer is."

What about the mines? Are they closed up for good, all the gold gone?

"They didn't mine all that ore out of those mines," he says with a gleam in his eye and an excitement in his voice that's contagious. The lure of gold is still an infectious disease around these parts.

When I left Dunton, Paul was sitting on the bench in front of the saloon, basking in the warm morning sun. He was burdened with a tough decision: Should he go for a soak in the hot pool, then run some errands over the pass in town? Or should he go to town first, then dip in the hot pool? I offered no advice.

Millo was sitting on the porch of a log cabin he was completing for himself, ready to put some finishing touches on it. He is no newcomer to the art of log cabin building, having built some of those in Dunton in the early days. This one is made of aspen logs, carefully hand notched (he laughed when I asked if he used a chain saw). It sits on a hillside across the river from "beautiful downtown Dunton," and commands a fine view of the river and aspen-gold mountainsides. It was so peaceful I wanted to delay my departure as long as possible. We made small talk about the approaching winter. He spoke of the elk bugling in the evenings, making me wish even more that I could spend another night. Finally I could delay no longer. I rose, we shook hands. As I left he was returning to his task of finishing his lovely home, built to last a century or two. Interesting, I thought, how a man 70 years old would put such great labor—and love—into building a new home for himself. He must plan on being around here for a long time to come.

• • • •

There was a time when we lived a lot closer to rivers. Or, at least some of us did. But the Mississippi riverboats of Mark Twain's era are gone now, along with the mountain men and the *voyageurs*, the birch bark canoe and the keel boat. Curiously enough, however, the last two decades have seen the emergence of a new breed of river people, ones who quite literally *live* rivers. Their names are Jim Campbell, Ken Sleight, Ron and Sheila Smith, Hank and Sharon Miller, Cort Conley, Lute Jerstad, Rick Petrillo, Martin Litton, and maybe a few dozen others. They call themselves guides or outfitters or, more simply, boatmen. Their businesses—or, in some cases, nonbusinesses—are an informal collection of rafts and trucks and people. Under such names as Grand Canyon Dories or Wonderland Expeditions or Wild Rivers Idaho, they run their inflated tubes of neoprene down the wild waters of the Colorado, the Green, the Selway, the Salmon, or the Snake. It would be very easy to assign purely monetary motives to their efforts, but it would also be very wrong to do so. True enough that Colorado River raft trips have become big business in recent years, but not all of these people capitalize on it; and many of them could care less about ledger books so long as they make expenses and don't starve in the off-season. They do what they do for very special reasons.

Like Jim Campbell. By all rights he should be termed a traitor. How else can you describe a man who refuses to embrace all the things worshipped so dearly by our society? Things like clocks and schedules, desks and deadlines, meetings, salaries, pensions, promotions, mortgages, status, and the rest? Why, the man doesn't even own a television set! I mean, what can be said for a shiftless lout who runs rivers for a living?

Well, a lot. For one thing, Campbell is a dropout from that old 9-to-5 routine himself, giving up a promising career as a nuclear physicist for the Atomic Energy Commission. Or, as he describes his former profession, "trafficking in neutrons, radiation, and other assorted and unhealthy goodies." Although his background is unique, Jim is rather typical of a number of people across the country—and especially here in the Rockies—who make a living as wild river guides. Taking people on such expeditions as a profession, according to Jim, is like being a tour guide in a museum, an instructor in the arts, a professor of philosophy, a cook, sea captain, geologist, ecologist, and psychiatrist all rolled into one.

River running has become a unique American phenomenon. It's difficult to say exactly how many people make wild river trips nowadays, but some estimates run as high as 100,000 a year, and most of that takes place here in the Rockies. In Grand Teton National Park about 45,000 people a season make the one-day float trip on the Snake River. And the

Denver Public Library, Western History Department

Lower Falls
of the Yellowstone.
Photograph by
William Henry Jackson,
1872.

Lower Falls of
the Yellowstone River,
Yellowstone National Park,
Wyoming

National Park Service puts the number of Colorado River trips through the Grand Canyon at about 15,000 per year. On the Green, the numbers total some 17,000, and for the Middle Fork of the Salmon, about 5,000. By the time it's all totaled, the numbers become significant. So much so, in fact, that there are now growing problems of human impact on these wilderness environments.

Why have wild river trips become so popular? "Escape," says Campbell. "And a bit of danger. It used to be that people could find some solitude and escape by regular camping. I remember when I first moved to the Rockies you could go to almost any Forest Service campground and have it pretty much to yourself. No more. Now, every campground is full, and some of 'em have become small-scale urban environments that people thought they left behind, with big motor homes and motor scooters, noise and squalor and even crime. And so more and more people are turning to things like backpacking and river running to really find some solitude."

Escape and solitude I can understand, but why danger?

"Well, I have a theory," says Jim. "People lead such humdrum, protected lives nowadays that they find themselves wanting, even *needing,* a little excitement and danger. The river provides it."

Jim averages some 400 people a year on his trips. And though his business is thriving, he uses much of his profit to hire the best people he can to work for him. Not just good boatmen, but those special people who bring an added dimension to the experience for others. Because Jim is also a subversive. Ever since he moved west, and long before starting in the river running business, he's been involved in trying to save some of this wilderness. The river trips, then, become a way for him and the others to sensitize people, make them aware of the values of wild rivers and wilderness country. Thus, Jim and his boatmen feel that they're in the education business, and most participants come away from these trips with profound changes in their lives.

Cort Conley is one of Jim's employees. He has a Doctor of Law degree from the University of California and enough offers from various law firms to make it big in the field, if he chose to. Instead, he prefers running rivers and in the off-season he travels—to Europe, Asia, South America, Africa, New Zealand. Conley has backpacked in over 40 countries around the world, and to broaden his own perspective he has worked as a forest ranger, logger, miner, roughneck on an oil crew, and packer in the High Sierra. He's also a certified sky diver and skin diver, and serves as chairman of the Idaho State Historical Society's oral history program. Jim Campbell describes him as a walking, one-man university.

"One of the most rewarding things about this business," says Conley, "is seeing the effect of wilderness and wild rivers on people. For some, their whole lives become changed by the experience. Every boatman has a special box of letters he keeps, letters from people telling how they got back home and quit jobs they've hated for years to start into something they've always wanted to do but never quite had the courage to try. It's a nice feeling to be a part of that kind of process. Interestingly, these people attribute their new insight and changes in their lives to the boatmen they met on these trips, but it wasn't really the boatmen at all. It was the river. And the wilderness. Getting away from all the pressures that surround them enabled them to think clearly about themselves and their lives, probably for the first time."

But Conley underplays the effect that he and others have on people making these trips. Campbell's business could almost be termed a floating university, a modern-day equivalent of that "Rocky Mountain College" described by mountain man Osborne Russell. Besides Conley and all of his amazing background, Campbell has working for him a published poet, an English professor, a PhD mechanical engineer, a chemical engineer, and a writer, all of them rabid environmentalists to boot. Thus, taking one of Jim's trips is like signing up for graduate studies in several of the arts and sciences. And a number of other river outfitters are the same.

Beneath the gruff exterior of Martin Litton, for example, is an amazing storehouse of knowledge of the history, archeology, and geology of the Colorado River country. It is hard to come away from a trip with Martin and not be affected by the man. A former editor of *Sunset Magazine* and a published writer and photographer in his own right, Litton was one of those who led the successful battle against the damming of the Grand Canyon several years ago. Today he conducts a rather unique kind of river trip, running the Colorado and other rivers in wooden dories somewhat reminiscent of the wooden boats used by John Wesley Powell. Litton is, in fact, an expert on Powell, and the highlight of many a trip with him is to drift along beneath towering sandstone walls while Litton recites from memory appropriate passages from Powell's journal.

Then there's Ken Sleight. Few people know the secret places of the slickrock country of Utah as well as he. Ken has led a battle for years to protect the spectacular Escalante Canyon from destructive developments (thus far successfully). I first met him over ten years ago when he and a friend drove their beat-up pickup truck all night from Salt Lake City to Boise in order to testify at a federal hearing on preserving part of the upper Selway River. It's the kind of thing that he does in his spare time when he's not running the Green or Yampa or poking around in Dark Canyon or Grand Gulch or his beloved Escalante River country.

Among river runners, as among most other professions, there are those people held in special esteem by their colleagues. Rick Petrillo is of that breed. A boatman's boatman, praised and envied for his special skill. Having traveled the equivalent of once around the world on white water, there is nothing about rivers that Rick does not know. Rivers, in fact, are an integral part of his life and one gets the distinct impression that his whole system, mental and physical, is attuned to flowing water. Between trips Rick is like a caged animal, restless, moody, even morose at times, feeling hemmed in by the trappings of civilization. But once underway on the swift current of the Salmon or Colorado or Owyhee, an instant transformation takes place. He becomes relaxed, confident, and happy, comforted by the surroundings he loves best. He talks about the river and at times he even talks *to* the river. Running white water with Rick at the oars is almost a mystical experience, like feeling firsthand all the superbly executed grace and skill of a slalom skier.

Rick and all the others involved in this wild and crazy profession are travelers along the twilight edges of our civilization, a unique breed poking into the last untouched remnants of a world most people are totally ignorant of. In his book of poems, *Catching it Whole,* Rick expresses it thus:

> *I go about self-appointed duties*
> *Afoot or in the pickup*
> *Driving cruelly rutted paths,*
> *Tracing out the boundary lines*
> *That corral men the length of their lives.*

The cruelly rutted paths and the wild rivers diminish in the onslaught of our civilization. Jim Campbell and the others see increasing problems for the future, even on those rivers to which we've assigned some measure of protection. What it boils down to is the fact that only a few sizable wild rivers remain, even in the Rockies, and the greater numbers of people seeking the beauty and adventure of these rivers are themselves creating increasing problems. The Colorado River typifies the worst situation, though most of the other rivers are not far behind.

In recent years some 15,000 people a year have been running the 270-mile stretch of the Colorado through Marble Canyon and Grand Canyon. The vastness of this wilderness would seem to preclude much impact. However, the use is confined to the narrow corridor of the river, a zone averaging probably only 300 feet in width along the 270-mile length of the trip. Thus, when one multiplies it out, the actual area of impact is roughly 14 square miles, which translates into a transient population density of a little over 1,000 people per square mile, not far from the average of 1,400 people per square mile of Los Angeles. True, this is a transient population and spread out over a four- to five-month season. The main point is, however, that this kind of continued heavy use year after year is bound to have severe consequences.

Overleaf:
Grand Canyon
National Park,
Arizona

In order to prevent the deterioration of our wild rivers, Campbell and others have been trying for years to have limits and quotas set on such rivers as the Snake in Hells Canyon and to ban the use of motors, which sharply increase the impact on wilderness. Campbell has already served notice that he will pull out, cease his commercial trips on the Snake, if it will spur the Forest Service into protecting the river. In talking of the future, Jim foresees the day when he may have to quit running other rivers as well, simply because his conscience will not allow him to contribute to further degradation of the wilderness.

In preparation for that time, Jim and another physicist, Warren Nyer, have purchased the remote Shepp Ranch along the Salmon River. Rather than the typical dude ranch operation, they talk excitedly of establishing an environmental learning center there, holding classes and seminars and conferences on subjects as wide-ranging as wildlife, writing, painting, music, alternative life styles, and energy systems. The wilderness environment of Shepp Ranch ("Twenty miles by trail," says Jim, "or four days by raft.") makes it an ideal campus for environmental studies. The ecosystems range from the river itself to alpine tundra in the mountains above. And as I sat with Jim in early spring discussing these plans, bighorn sheep and elk grazed, literally, in his front yard.

"And if it doesn't work," says Jim, "I'll just prop my feet up on the porch rail and spend a few decades watching the Salmon River roll by." But one has the feeling that Jim and his people will make it work, however.

·   ·   ·   ·

River Runners. Boatmen. Guides. Call them what you will, they represent an old tradition of the free spirit, of independence and adventure. They are the young version of an old breed, cohorts of the Cowboy, the '49er, the Oregon Immigrant, the Mountain Man.

One of Rick Petrillo's favorite poems is not one of his own writing, but very well could have been. It was penned by Badger Clark, cowboy poet, entitled "The Passing of the Trail":

> The trail's a lane, the trail's a lane.
> Dead is the branding fire.
> The prairies wild are tame and mild
> All close-corralled with wire.
> The sunburnt demigods who ranged
> And laughed and loved so free
> Have topped the last divide, or changed
> To men like you and me.

For a while, at least, those sunburnt demigods are alive and well on some of the rivers of the Rockies.

The Great Dream:
The Blooming
of the West

PARADOX: IT IS A LAND of snowcapped mountains, lovely alpine lakes, and great rivers. At the same time it is largely arid, averaging only 16 inches of rain a year over the whole geographic region, a place called an "oasis civilization" by historian Walter Prescott Webb.

But if so many big rivers have their origin in the Rockies, how can the land be arid? To answer that question, one must look closely at the topography and climate of the region. The Rockies, as pointed out earlier, are not a single unified chain of mountains, but a series of unconnected ranges. Such mountain ranges as the Lemhi and the Beaverhead, or the Absaroka and the Bighorn, the Wind River and the Medicine Bow are separated by large areas of prairie or desert; in a sense one might consider the high mountain country to be a series of islands in a sea of arid plains. The islands receive a great deal of precipitation when the moist westerly air is raised and chilled by the cold peaks to release its moisture. The plains, by contrast, receive relatively little of this atmospheric moisture. In the mountains precipitation occurs primarily in winter, and the snow piles up deeply, then melts slowly in spring and summer to feed the rivers. The mountains become, then, something akin to a bank vault, a repository for these aqueous savings. The rivers might be considered the dividends, but they are dividends that are paid very slowly. Thus, what one finds here in the Rockies is that the flat and gentle lands that might be suitable for cultivation and settlement are without the necessary water. And the mountains, harsh, rugged, uninhabitable, have water aplenty.

Such is the way of life here in the Rockies. It is dry, very dry. Relative humidity here in the Front Range of Colorado, for example, rarely exceeds 30 percent. In many other places it is far lower than that. Living in the super-dry Snake River Plain, we used to joke about stapling postage stamps to letters because it was so hard to work up enough moisture to lick 'em.

I remember how it used to drizzle (and probably still does) for days, even weeks on end back in New England. I recall many a two-week vacation when the sun never showed once. Well, maybe once—on the way home. But here in the West, "drizzle" is an alien term. When it rains here, it is a brief thunderstorm. Even then, over in the really dry country of Utah and Arizona and Nevada, some of the precipitation will evaporate before it ever reaches the ground.

All of this is pretty obvious when one travels throughout the Rocky Mountain West. There are vast, sage-covered plains where the only trees are those growing along watercourses. Forests cling to the moisture-giving mountain slopes or confine themselves to the

Marble Canyon,
Colorado River,
Arizona

Virgin River, Zion National Park, Utah

Canyon De Chelly,
Arizona. Note tents
at lower left.
Photograph by
Timothy H. O'Sullivan,
ca. 1870.

Denver Public Library, Western History Department

protected river valleys. And in summer one can plan a two-week vacation and expect
14 days of clear blue skies, punctuated only by an occasional storm that might last a
few hours.

It is unfortunate that the migration of our culture moved from the moister climes
of the East to the drier ones of the West. Because if the West had somehow been settled
first, perhaps we would have learned to adapt better to a dry-land ecosystem. Instead we
were spoiled by the wet and fertile lands east of the Mississippi and in moving west we
expected crops to spring from the ground and thrive. But they don't. They die. Unless—
and here's the key—unless you find ways to bring water to them in lieu of the nonexistent
rainfall.

In the wake of the throng of settlers migrating to Oregon and the hordes who sought
gold, there were many who saw this dry Rocky Mountain region as a land of plenty, a veritable Garden of Eden. Brushing aside those earlier reports of Pike and others who declared
this "The Great American Desert," people such as William Gilpin, first Governor of the
new territory of Colorado, spoke in glowing terms of the great potential for settlement and
growth in the Rockies. According to Gilpin, this resource-laden land could support a population in the hundreds of millions.

It was a strange time in American history. Perhaps as an escape from the realities
of strife brought on by a bloody Civil War, people were ready to believe anything. Even
one report that solemnly declared firsthand knowledge that the Colorado was a peaceful
and navigable river from the Colorado border to its mouth at the Gulf of California. (In a

few years the journey of John Wesley Powell would lay to rest that myth.) The politics of the time lead one to wonder whether the Homestead Act, signed into law in 1862 by President Lincoln, hadn't been enacted rather hastily and recklessly to speed up settlement of western territories and to quickly bring them into the Union as free states, thus strengthening the North.

Whatever the motivation, the fever of expansion brought settlers on the heels of the miners to the Rocky Mountain country. And it soon became apparent to most that 160 acres of free land under the Homestead Act were not capable of sustaining a man and his family in the dry, barren stretches of plain. Instead of farming, many turned to ranching, but even then water was still the key to survival. The prized lands were those settled along the watercourses, and many a bitter and bloody battle was fought over the rights to that water. Despite the hardships, lands were settled—some 250 million acres under the Homestead Act. In addition, railroads had been offered huge tracts of land as enticement to expand and link the continent from east to west. Thus the Rockies were rapidly being carved up.

It is within this framework that John Wesley Powell arrived on the scene in the late 1860s and, his voyage on the Green and Colorado rivers notwithstanding, made one of the greatest contributions in land politics as director of the United States Geological Survey. After his decade of exploration and scientific study of the resources, watercourses, climate, and native cultures of the Rocky Mountain West and Southwest, Powell published his *Report on the Lands of the Arid Region of the United States* in 1878. In light of the expansionist fever of the nation, it was a bombshell, described by Wallace Stegner

Canyon del Muerto,
Arizona.
Photograph by
Edward Curtis,
ca. 1885.

Green River near
confluence with Yampa,
Dinosaur National Monument,
Utah and Colorado

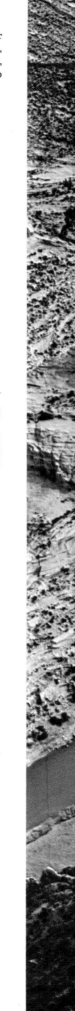

Petroglyphs near Snake River in Hells Canyon, Idaho and Oregon

as "a complete revolution in the system of land surveys, land policy . . . and farming methods in the West, a denial of almost every cherished fantasy and myth associated with the Westward migration and the American dream of the Garden of the World."

Powell called for recognition of the arid nature of the region and for dealing with human settlement in terms of that aridity. The haphazard and arbitrary nature of the Homestead Act did not take into account the availability or nonavailability of water. Powell's plan urged that the size and even the shape of land parcels be adjusted to conform to topographic, climatic, and riparian conditions. In other words, settlement and development of the region should be managed so that maximum benefit could be gained with minimal damage to the land. Perhaps more important, however, the plan set forth a philosophy for dealing with the region's rich resources. Powell recognized that the resources of the Rocky Mountains were of potential benefit to the nation as a whole and that all citizens had a stake in the forests, minerals, and waters of the region. He asked for careful conservation of these resources and for careful planning for their development. Like most great men, however, Powell was far ahead of his time, and much of what he proposed would not be implemented until long after his death in 1902. He did see passage of the Forest Reserve Act of 1891 setting aside some 13 million acres in the West as federal reserves. But it was Teddy Roosevelt, almost four years after Powell's death, who expanded that system to almost 170 million acres, establishing what became the National Forests in an attempt to halt the reckless timber barons. And while Yellowstone and Yosemite had been set aside as national parks by the 1870s, it wasn't until 1916 that the National Park Service was established and impetus given to preserving even more national treasures of lands and waters. And finally, Powell's reclamation schemes, by which certain waters of the West might be used to make the arid lands arable, did not begin to be fully implemented until well after he died.

Black Canyon
of the Colorado
(now flooded by
Lake Mead behind
Hoover Dam).
Photograph by
Timothy H. O'Sullivan,
1871.

Powell's work had laid the foundation for planned resource development on a regional scale; and in the early part of the 20th century, the implementation of such planning began to change the face of the West. Every major river of the Rockies was ultimately mapped for potential development. Dams were built, first for irrigation, then for hydroelectric power, then for both. Growing population centers as widely spread as Denver and Los Angeles laid claims to ever-increasing quantities of these flowing waters in order to quench the thirst of industry and people. The land bloomed in places, thanks to the Bureau of Reclamation. Floods were halted in places, thanks to the Army Corps of Engineers. And cheap electrical power became available, thanks to both. Dams, water diversions, irrigation schemes, power plants, crops, growth, and progress. It all seemed to fulfill Powell's hopes and plans far beyond his wildest dreams. But one wonders if, in visiting and studying the West today, Powell might not be seriously troubled.

. . . .

In the last decade or two, the Rocky Mountain West has become the focal point of a growth and development boom unparalleled in its own history and possibly without peer in the nation's. At the root of this growth are two driving forces: unspoiled lands and water and air that seem to offer people an escape from the environmental horrors elsewhere; and vast deposits of coal and oil shale that are touted as the solutions to the nation's energy problems. The first is resulting in an incredible influx of people seeking new jobs and new lives in the area's urban centers. An average of 1,200 new people a week—or one every nine minutes, as the chamber of commerce likes to brag—come to Denver to stay. The second force is concentrating construction workers, miners, and laborers in the rural areas in and around those coal and shale resources, the heart of the wide open spaces. This boom in both urban and rural population promises to strain the West's most critical resource—water—as it has never been strained before.

The unprecedented rapidity and scope of this growth make planned resource development not merely an admirable goal, but a necessity. While "progress" has for generations been the quintessential American dream, we have learned the hard way that growth and development do not always represent progress. The interdependence of living things requires that the long-range effects of proposed development schemes be carefully considered and that such projects as are undertaken be managed in a way that will minimize their impact on the environment. While we are still far from finding solutions to many of the problems related to resource development, several states—not only in the West—have already imposed tougher air and water standards as well as more stringent laws to govern mining reclamation.

It is clear that the challenges confronting the Rocky Mountain states today are of concern to everyone in the nation. The region is repository for many, if not most, of the nation's wilderness areas, national parks, and wild rivers. These unspoiled preserves become more valuable to us as time goes on. If nothing else, they serve to remind us of what it was once like over much of our land and might be like again if we work at it. If the region is altered, destroyed, then we all lose the opportunity for even brief escape from our crowded, complex civilization, the chance to sit by a wild river and learn something of the wonders of the natural world.

Colorado River, Grand Canyon, Arizona

Living River

WATER IS AN ALIEN HABITAT to us. Oh, we *seem* familiar enough with it, but except for an occasional trout or bass, our contacts with the inhabitants of this ecosystem are rather limited. Moreover, our vague understanding of the water biome seems confined to still water—lakes and ponds. Somehow, it's difficult to picture the life of moving water.

But life there is and plenty of it. From their very birth our streams and rivers teem with living things, plants and animals that have adapted to flowing water. In fact, even the raindrop that feeds the streams may contain particles of pollen and dust and dissolved gases, living and nonliving nutrients for other life forms. It seems that there is no such thing as pure water in nature. Those sparkling mountain streams carry dissolved minerals, particulate matter, and microscopic organisms, all of which form an integral part of the web of river life.

One of the simplest life chains related to streams begins up in the alpine tundra, where hikers and mountain climbers often puzzle over a strange red stain on the surface of snowfields, what we used to call "watermelon snow" in the northern Rockies. It is the result of red algae that exist on the surface and survive in that narrow temperature range very close to freezing. These algae, together with windblown pollen, are the food source of the glacierflea, a tiny relative of the silverfish, and the glacierflea is the exclusive food of the harvest-man spider. And finally, completing this life chain, the harvest-man spider is the prey of several species of birds that dwell temporarily in the high country during the brief alpine summer.

As the snow and ice melt, the trickles of water pick up tiny bits of mineral matter and decaying vegetation, both of which are nutrients for more life. As these waters join, then gather momentum, they provide a habitat for various insect larvae, snails and worms, algae and mosses. Obviously these plants and animals must be adapted to fast-flowing water. In fact, one alga, which grows as a thin film on submerged rocks, is so adapted to fast water that it will not survive in slower waters. (It is this alga, by the way, that makes streambed rocks so slick and treacherous to someone wading across a mountain stream.) Many of the insect larvae of swift water have hooks and pincers and other mechanisms for holding fast, whereas their counterparts in slow or still water are less highly adapted. The caddis fly and mayfly are excellent examples of effective adaptation. Both spend their nymph or larval stage in fast-moving streams. The caddis fly nymph builds a protective case of tiny pebbles cemented together with salival secretions, occasionally attaching larger stones to act as ballast against the current. The mayfly has a flattened body and numerous

protruding hooks enabling it to sneak into cracks and crevices and hold on against the rush
of water. The emergent adults of both, delicate-winged and graceful flying insects, are important food for trout.

Trout. No species, except perhaps the salmon, more typifies the Rocky Mountain
streams and rivers. They find their ideal habitat here, for they require cold, clear, swift,
and well oxygenated water for survival. Rainbow, cutthroat, golden, Dolly Varden,
brookies. Even I, no ardent fisherman, must admit to a great deal of excitement on the few
occasions when I toss a line into a mountain stream. Facing upstream and fixed in one spot
by slow, subtle movements of its body and fins, a trout lies in wait for something to come
its way. This may be a mayfly or a fisherman's dry fly; but when it floats near, the trout
explodes into action, sometimes breaking totally out of the water in its surge to devour
the prey. And then (if it's not hooked by that dry fly) it moves swiftly back into position to
await something else that might come its way.

As the mountain streams continue their descent, they may encounter a variety of
conditions that determine the kind of life forms which can flourish in their waters. The
stream may plunge steeply over rocks and cliffs, and even under these torrential conditions certain species may thrive, nurtured by spray and mist. Or the waters may suddenly
enter a quiet mountain lake formed by the damming effect of a glacial moraine. These still
waters may support new insects and plants, different strains of algae, certain grasses and
mosses, and occasionally frogs and snakes if the lake is not above the survival zone for am-
phibians and reptiles. Fed by nutrients carried by inlet streams and sustained by the oxygenated water of those streams, such lakes are home for many varied species, sometimes even
well above timberline. Then the waters move on, plunging downward and joining together
with more streams, all to become a river somewhere.

Question: When does a stream become a river? On some maps the last "r" of the
name "Colorado River" rests on the Continental Divide. Obviously it is not a river at all
at this elevation, but a stream. Or, more correctly, a trickle. Semantics aside, there is some
point at which all of these mountain waters have coalesced in lowland valleys and can be
called a river. Here, with lesser speed and larger volume, begins a new habitat. And it is
here that the impact of man upon the river most often begins.

Shoshone Falls,
Snake River, Idaho.
Photograph by
Timothy H. O'Sullivan,
1874.

Snake River, Swan Valley, Idaho

Prickly pear
cactus

No two rivers are identical in physical, chemical, or biological conditions. Before it was dammed extensively, the middle and lower part of the Colorado River ran thick and muddy, carrying a heavy burden of silt and sand, the erosion products of the sandstone country and the outwash of the Rockies. "Too thick to drink, too thin to plow," was the old saying. But the river life had adapted to the load of sediment. Not the same kind of life found in or by the Salmon or the Clearwater, perhaps, but these life forms became indigenous to the Colorado. Now, of course, many of our rivers are changing. Because of the many silt-collecting dams, the Colorado runs clearer than before. Trout now replace the channel catfish, but no one can say for sure what long-range effects this change will have.

One species whose life cycle requires suitable conditions throughout its river habitat is the salmon. Born in the swift, shallow upper tributaries of the Pacific drainage of the northern Rockies, Chinook salmon spend the first several weeks of life feeding in these cold, clear waters. Then, triggered by one of those strange biochemical changes, the fingerlings begin a slow migration downstream, allowing the swift current to bear them along, tail first, as they continue to feed and grow in the ever-deepening waters. The journey, if it begins in the upper reaches of the Salmon River in the Sawtooth Mountains, may involve a total distance of almost a thousand miles, first down the length of the Salmon, then the Snake River, and finally into the Columbia and on to the Pacific Ocean. In making this migration, the salmon must adapt from the food sources of the swift mountain stream to those of the large, slow-moving, and somewhat warmer rivers. But the most abrupt change occurs when they enter the ocean and become a seagoing species. As a part of the ocean life chain, the young salmon head north toward Alaskan waters, feeding on the rich plankton in the sweeping arc of the Japanese current. In turn they become prey themselves to other species such as sea bass, sharks, sea lions, loons, ducks, and grebes. As they grow from immature smolt to adult size, the salmon become predacious themselves, feeding on smaller species. Then, after some four years of ocean life, another biochemical change triggers that strange homing and breeding instinct, and the salmon head south, toward those river waters from which they entered the sea.

After five years of ocean life, a Chinook, or king, salmon averages 20 to 30 pounds in weight as it enters the mouth of the Columbia and begins the last great journey of its life. It has now become a mobile and super-sensitive analytical chemistry laboratory, detecting those minute chemical "odors" unique to its original home. Through swift currents, rapids, and waterfalls, it battles its way upstream, sampling the chemical content of the waters instinctively. At the mouth of the Snake River, the odor becomes richer, and the salmon turns here to follow these waters. Then a pause at the mouth of the Salmon, and again the scent becomes stronger. Mineral content, nutrients, perhaps even certain dissolved gases are detected by that keen sensory system, and the salmon rejects tributary after tributary, fighting upstream toward its goal: the very gravel spawning bed where it was born. Finally, if it is one of the lucky survivors of that arduous journey, it arrives in the swift upper Salmon River. Here a nest is scooped out in the gravel bottom and the eggs laid and fertilized. Then these magnificent fish weaken rapidly and within a few days they die, their battered bodies littering the shore of the waters that gave them birth. But the life cycle has been completed, and soon new fingerlings will start the journey over again.

Like most other river life, the salmon is reasonably tolerant of minor changes to its habitat. There have always been certain natural forms of pollution. Landslides and earthquakes alter mineral content and increase sediment, while forest fires radically change watersheds and speed runoff. Decaying organic material, from leaves to animal carcasses, has been an integral part of the river's life chain. The catastrophic changes are generally short-lived, while minor variations in organic "pollutants" are easily adapted to, for rivers have an amazing capacity to purify themselves. Tumbling waters become enriched in oxygen, the best purifying agent, and bacteria and algae cleanse the river of certain organic matter. But there are limits to the alterations that a river—and its attendant life forms —can adjust to or overcome, and in a growing number of rivers of the Rockies we are already crowding those limits.

In recent years, we have all had called to our attention some of the effects that man-made pollutants can have on aquatic life. But the introduction of chemical wastes into water is not the only way man can influence river ecology. The far-reaching and often unforeseen results of damming are a case in point. In the first place, the huge reservoir of water trapped behind a dam presents a large surface area to absorb solar energy, thus increasing the temperature of the river system downstream. Most of the life forms dependent on cold, fast-moving water are replaced by species adaptable to this new, slack water habitat. Then there is the effect of drawdown, that is, the lowering of the water level in the reservoir. Unlike the normal cycle of seasonal flooding of a river during spring runoff, drawdown can cause unseasonal and irregular changes in the water flow. For example, on certain portions of the Snake River where Canada geese normally nest, there is an unusually high mortality rate of nests and young because the birds, attuned to seasonal cycles, build their nests at a time of year after the normal peak of runoff. But irregular releases from the reservoir to meet power demand or irrigation needs may raise the river level at almost any time and flood out the nesting birds. Dams can also be a barrier to migrating species such as salmon. Some attempts have been made to minimize this effect. Fish ladders presumably allow migrating fish to bypass these concrete barriers, but in some instances the salmon milling around at the base of a spillway have to be trapped and trucked around the dam. Finally, a series of dams along a single river system can alter the very chemistry of the waters, particularly in dissolved gas content. Water pouring over the many spillways traps air and forces it into solution. The amount of dissolved nitrogen may be increased manyfold over normal conditions, subjecting the river's inhabitants to nitrogen narcosis, something akin to the deep-sea diver's malady, the bends.

• • • •

Salmon fishing,
Columbia River.
Photograph by
Edward Curtis,
ca. 1890.

Medano Creek and Sangre de Cristo Range, Great Sand Dunes National Monument, Colorado

Great Sand Dunes National Monument, Colorado

Forests of the Rocky Mountains play an important role in river ecology. Grasses, shrubs, and trees on an undisturbed mountainside form an almost continuous cover of vegetation, a mantle that is able to absorb huge quantities of precipitation and then release it slowly into the web of water that feeds the rivers. When that vegetative cover is disturbed, as by logging or road building, the runoff is increased manyfold. There is growing evidence that damaging floods along some of the rivers of the Rockies can be traced to disturbances of the watersheds of these rivers. Clear-cutting of forests, for example, opens whole mountainsides to larger amounts of runoff, swelling streams to flood proportions after storms.

There is a close kinship here in the West between rivers and life, a dependence made all the more apparent by the ruggedness and harshness of the country. All life is dependent upon water, of course. In moister climes and gentler terrain, there is less need for life to cluster around large, primary sources of water such as streams and rivers and lakes. But in large areas of the Rocky Mountain West, both the density and the diversity of plants and animals diminish radically with the distance from the nearest river. As Powell noted on his Colorado River journey, the immediate vicinity of the river can be quite friendly, with willows, cottonwoods, oaks, and numerous other plants, plus birds and mammals to soften the harshness of the land. But leave the river, and the land is hostile, bristling, sparsely populated desert, a place where survival is measured by the ability to find water somehow, some way. And the sooner the better.

In the central and northern Rockies, certain animals congregate along rivers for another reason: winter. When the high country is given over to deep snows and howling winds, the big game herds migrate to the protected and milder lowland valleys and canyons. In Hells Canyon on the Snake River, for example, elk and deer begin migrating in late fall from the snowbound rims to river level where there is little snow during the winter. In fact, the contrasts in this gorge are quite amazing. On trips there in mid-April, a time of year still considered winter throughout most of the Rockies, I've found the grass lush and green, and flowers blooming along the river, while a few thousand feet above, the forests are still white with deep snows. And there seems to be a herd of deer or elk around every bend of the river.

In traveling on or beside the rivers of the Rockies, one can hardly fail to notice abundant bird life. The highlight of many a raft trip is the spotting of some of the raptors soaring above. The golden eagle, the rarer bald eagle, the extremely rare peregrine falcon (found now in only a few places), plus several other species of hawks are all dependent on river habitat. The osprey, or fish hawk, is still common along sections of the Snake River in Grand Teton National Park and in the Grand Canyon of the Yellowstone farther north. Unfortunately, it is the osprey and the eagle and certain of the falcons that have suffered from deterioration of the rivers. Concentration of DDT in river food chains reaches such high levels in some fish, prey of these birds, that the subsequent transfer of that toxin to the birds has affected their reproduction and hatching capabilities. With the ban on DDT, there are signs of an increase in the numbers of these species.

These river byways are important habitat for quite a diversity of bird species. For example, along the South Fork of the Snake River in eastern Idaho is found one of the largest heron rookeries in the region. In the spring when the young of the great blue heron are hatched, the rookery is a cacophony of noise. The huge, slate-gray parents return with freshly caught fish to feed the squalling youngsters in nests high in the willow and alder thickets along the shore. Everyone seems to vocalize loudly. Snowy egrets and black-crowned night herons nest in the same thicket and, whether screeching for space or fighting over food, there is rarely a moment of silence.

One could hardly end a discussion of birds and rivers without mention of the trumpeter swan. Once on the verge of extinction, these graceful and elegant birds have increased in numbers sufficiently in recent years to make their survival reasonably assured. Thus, today it is more and more frequent that one or more may be seen floating serenely in the Snake River in the Teton country or somewhere along the Yellowstone River in Yellowstone National Park.

Rivers are vital habitat for countless small mammals. Otters swim and fish the year around in nearly all the rivers of the Rockies. In fact, they give the distinct impression that they consider rivers the ultimate playground, using the grassy or snow-covered river banks as slides, and playing endless games in the cold, swift waters. I recall that on one trip on the Salmon River two young otters followed our raft, swimming in circles and diving as though inviting us to join their game; then they slipped out on shore to sit up and watch, puzzled and perhaps disappointed, as we drifted around a bend out of sight.

The mammal that derives most benefit from mountain streams is the beaver, the furred gold of the mountain man. Beaver play an interesting role in both stream and forest ecology. Their dams create networks of ponds which in turn become habitat for a number of other animals. In certain areas of the northern Rockies, for example, moose are frequently seen standing in the waters of a beaver pond feeding lazily on the aquatic plants that could only have been nurtured by such a pond. And numerous waterfowl find the quiet waters ideal for breeding and feeding. But perhaps the most far-reaching effect of beaver and their activities is the change brought about in nearby forests. The ponds created by the dams may frequently flood lowland forests, killing the trees and creating park-like openings in the forest. The standing dead snags then become favored perching places for the fish-hunting hawks and eagles and are frequent nesting spots for these and other species, such as woodpeckers. The favored food of beaver is aspen, especially the succulent, fresh inner bark and tender shoots, and whole groves of these trees may ultimately be cut down and dragged off for food (some of it stored to tide them over the winter) and for building and repairing dams. Eventually the aspen forest may give way to an open grassy meadow, providing more grazing and browsing area for elk and deer. Or the removal of an aspen grove may hasten the natural succession of such other species as spruce or fir. The beaver move on when their food supply has dwindled. Slowly, over a period as long as 30 to 40 years, the forest once again invades the meadow; and the stream, having washed away the abandoned dams, now flows free again, edged once more with willows and spruce and aspen. Sometime later a new family of beaver may decide that this is ideal habitat, and the cycle starts again.

·   ·   ·   ·

Here in the West one hears the phrase "working river" with great frequency. The implication is that unless the waters are put to use driving turbines or watering crops or carrying wastes, they serve no purpose. This failure to recognize that rivers are an integral strand in the web of life continues even today in many parts of the Rockies. But there is a glimmer of hope: There seems to be growing general awareness of alternatives that might spare resources and save the rivers, though we've been slow in acting on those alternatives. We still have a way to go, however, before becoming attuned to the philosophy expressed in 1854 by the chief of the Suquamish tribe of Washington Territory:

> The rivers are our brothers. They quench our thirst. The rivers carry our canoes,
> and feed our children. If we sell you our land, you must remember, and teach
> your children, that the rivers are our brothers, and yours, and you must hence-
> forth give the rivers the kindness you would give any brother.

Chief Seattle

Old cabin and hay lifter

Storm over South Park, Colorado

A Brief Note on the Photographs

SOME PEOPLE HAVE IT MADE. Like me, for example. As I wandered the length and breadth of the Rockies shooting the pictures for this book, I was in many instances following in the footsteps (quite literally, at times) of some pretty hardy pioneer photographers. Men like William Henry Jackson, Timothy H. O'Sullivan, Edward Curtis, E. O. Beaman, Jack Hillers, and others. I've been a fan of some of these men for many years, and I'm also aware of the incredible work and dedication demanded of them in compiling their superb photographic record of the West back in the days when it really was the *wild* West. Practically all of them used the wet-plate process, which required coating glass plates (some of them up to 11"×14" in size) with emulsion, making the exposure, and developing this delicate negative, *all done right on the spot.* They roamed from canyon bottom to mountaintop—and on occasion floated a wild river or two—with bulky equipment that weighed 300 pounds or more.

As I said, I've been lucky. Not long ago I stood on a mountaintop in the western Tetons at the very place where William Henry Jackson made the famous first photographs of Grand Teton in 1872. It took him two pack mules to get his equipment up there. For my part, I simply hiked up with a light rucksack of camera equipment and film, equipment that would have made Jackson extremely envious.

Almost all my work is done in the 2¼"×2¼" format, and most of the pictures in this book were made with a Rolleiflex SL66. I never keep a record of technical data on any of my pictures because I feel it's unimportant. Let's just say that these pictures were made with lenses of various focal lengths, ranging from 40 mm. and ultrawide-angle to 500 mm. telephoto. Film was almost exclusively Ektachrome-X, with an occasional roll of High Speed Ektachrome used where lighting conditions required.

Even though it was done without all of these blessings of modern technology, the work of Jackson and Curtis and O'Sullivan and the others still stands as some of the finest ever done in the Rocky Mountains.